人工智能开发与实战丛书

PyTorch 深度学习实战：
构建、训练和部署神经网络模型
（原书第 2 版）

［印度］普拉迪帕塔·米什拉（Pradeepta Mishra） 著

王文通 刘 强 译

机械工业出版社

本书提供使用PyTorch开发深度学习应用程序的基本原理和方法，旨在为读者介绍机器学习工程师和数据科学家在解决深度学习问题中所采用的主流现代算法与技术，紧跟深度学习领域的最新发展趋势，助力初学者熟练掌握PyTorch。本书的核心优势在于，采用易于理解的问题与解决方案的结构，全面而详尽地讲解了PyTorch的使用方法，并提供了大量相应的代码示例，以便将这些概念顺利应用于实际项目中。

本书适合对计算机视觉、自然语言处理等领域感兴趣的人士阅读。对于希望在深度学习任务中运用PyTorch的读者，本书将是一本实用的指南。

First published in English under the title
PyTorch Recipes: A Problem-Solution Approach to Build, Train and Deploy Neural Network Models (2nd Ed.)
by Pradeepta Mishra
Copyright © Pradeepta Mishra, 2023

This edition has been translated and published under licence from APress Media, LLC, part of Springer Nature.

此版本仅限在中国大陆地区（不包括香港、澳门特别行政区及台湾地区）销售。
未经出版者书面许可，不得以任何方式抄袭、复制或节录本书中的任何部分。
北京市版权局著作权合同登记　图字：01-2023-3249号

图书在版编目（CIP）数据

PyTorch深度学习实战：构建、训练和部署神经网络模型：原书第2版 /（印）普拉迪帕塔・米什拉（Pradeepta Mishra）著；王文通，刘强译. —北京：机械工业出版社，2024.8

（人工智能开发与实战丛书）

书名原文：PyTorch Recipes: A Problem-Solution Approach to Build, Train and Deploy Neural Network Models, Second Edition

ISBN 978-7-111-75919-5

Ⅰ.①P… Ⅱ.①普… ②王… ③刘… Ⅲ.①机器学习 Ⅳ.①TP181

中国国家版本馆CIP数据核字（2024）第106334号

机械工业出版社（北京市百万庄大街22号　邮政编码100037）
策划编辑：杨　琼　　　　　责任编辑：杨　琼
责任校对：贾海霞　李　杉　　封面设计：马精明
责任印制：单爱军
北京虎彩文化传播有限公司印刷
2024年8月第1版第1次印刷
169mm×239mm・18.25印张・314千字
标准书号：ISBN 978-7-111-75919-5
定价：109.00元

电话服务　　　　　　　　　网络服务
客服电话：010-88361066　　机　工　官　网：www.cmpbook.com
　　　　　010-88379833　　机　工　官　博：weibo.com/cmp1952
　　　　　010-68326294　　金　书　网：www.golden-book.com
封底无防伪标均为盗版　　机工教育服务网：www.cmpedu.com

译 者 序

近年来，人工智能产品与解决方案的开发已逐渐成为主流趋势，因此基于图计算框架的需求正不断攀升。本书的核心主题是利用 PyTorch 框架开展深度学习建模。作为当前人工智能领域的热点技术，深度学习在计算机视觉、自然语言处理等领域具有广泛应用。

本书以问题与解决方案的形式，介绍了在 PyTorch 中构建、训练和部署各类深度学习模型的技巧，并提供了大量相应的代码示例，以便将这些概念顺利应用于实际项目中。本书的核心优势在于，采用易于理解的问题与解决方法的结构，全面而详尽地讲解 PyTorch 的使用方法，助力读者迅速掌握 PyTorch 在实际项目中的应用。本书适宜初学者以及具备一定 PyTorch 经验的技术人员阅读，目标是让读者学会运用 PyTorch 解决实际问题，成为熟练的 PyTorch 用户和实践者。

本书的具体内容如下：

第 1 章介绍了 PyTorch 的概述及其核心模块，同时还介绍了机器学习和深度学习的理念。这为读者对 PyTorch 框架及神经网络的全面理解奠定了基础，为后续章节中利用 PyTorch 构建深度学习模型奠定了基础。对于具有 Python 和 PyTorch 先验知识的读者，本章为复习内容。对于 PyTorch 框架的初学者，掌握这些基础知识至关重要。在深入探讨高级主题之前，熟悉相关术语和基本语法至关重要。

第 2 章介绍了采样分布及其生成随机数的方法。神经网络的核心是关注张量运算。在机器学习或深度学习模型的实现过程中，均需进行梯度计算、权重更新、偏差计算并持续调整偏差。本章还介绍了 PyTorch 所支持的各类统计分布及其适用场景。

第 3 章介绍了如何运用 PyTorch API 构建简单的神经网络模型，并通过调整超参数（如学习率、迭代次数和梯度下降方法）来优化模型参数。在研究卷积神经网络和循环神经网络的示例过程中，为这些网络引入了 Dropout 率以控制过拟合现象。

第 4 章介绍了各类激活函数及其在不同场景下的应用策略。在挑选最优激活函数的过程中，应依据准确性作为衡量标准；在模型中应动态选用能带来最佳结果的激活函数。

第 5 章介绍了监督学习算法的两大主要类别，即线性回归与逻辑回归，并分别介绍了在 PyTorch 中的实现方法。这两种算法均为线性模型，前者旨在预测实际值输出，后者则旨在将某一类别与另一类别区分开来。

第 6 章介绍了利用各类方法从训练数据集中训练深度学习模型以学习参数的过程。尽管卷积神经网络、循环神经网络及深度神经网络在架构上存在差异，但它们的训练过程以及超参数的选择方法却具有相似性。恰当地选择超参数对于提高训练效率至关重要，通过调整超参数，可以使得训练过程更为高效。

第 7 章介绍了如何运用词嵌入、连续词袋模型以及构建长短时记忆网络的实例。这些实例所对应的 PyTorch 函数可用于搭建自然语言处理流程，进而开发诸如文本分类、自动文本摘要、情感分析等诸多与自然语言处理相关的解决方案。

第 8 章介绍了在 GPU 环境中运用分布式 PyTorch 以及以并行方式处理模型训练的实践案例。另外，还展示了如何将大型深度学习模型压缩为较小尺寸，同时确保准确性的实例。量化策略在缩短深度学习模型推理生成时间方面具有重要意义。

第 9 章介绍了图像与音频领域的数据增强技术，涵盖了波形变换、图像滤波以及增强等。同时，还介绍了运用 PyTorch 进行特征工程以及提取相关特征的方法。

第 10 章介绍了 Captum 库和 Skorch 库的相关知识。Captum 为在 PyTorch 框架中开发的深度学习模型提供了运行模型可解释性的途径。而 Skorch 则允许在 PyTorch 模型上运用 Scikit-learn 函数和 API，如管道、网格搜索和交叉验证等。

本书内容覆盖面广。大部分算法均提供简洁且现成的 Python 源代码，便于读者进行验证。在此基础上，稍作修改和扩展，便可打造出适用于个人需求的项目代码。本书重点关注算法结论的分析与应用，使读者能够迅速掌握各种算法的特点及使用方法，而非纠结于算法细节。此外，本书结合实例，丰富实用，深入浅出地讲解了神经网络模型中典型且广泛应用的算法。

本书的翻译工作安排如下：王文通负责文前部分和第 1~10 章的翻译工作；刘强参与部分内容的翻译工作以及承担全书的统稿及审校任务。

本书的出版得到了北京市教育委员会项目（22019821001）、北京石油化工学院致远科研基金项目（2023014）的资助，在此表示衷心的感谢。最后，真诚感谢编辑杨琼为本书出版所付出的努力。

由于水平有限，译文中难免有不当之处，敬请各位读者批评指正。

关于作者

Pradeepta Mishra 是一位在人工智能领域具有深厚造诣的专家，在数据科学和人工智能架构方面拥有丰富的实践经验，目前担任 FOSFOR 公司旗下 5 个产品的自然语言处理、机器学习及人工智能计划的主管。FOSFOR 公司作为业界领先的人工智能和机器学习创新公司，始终站在技术前沿。Pradeepta Mishra 的专长在于设计卓越的人工智能系统，尤其在理解和处理自然语言以及基于自然语言处理的推荐系统方面展现出超凡的专业能力。

他作为发明人共申请了 12 项专利，并在学术领域取得了丰硕的成果。他先后撰写了 5 本著作，内容涵盖数据挖掘、空间数据、文本数据、网络数据和社交媒体数据等多个领域。这些书籍均由知名出版社出版发行，分别为《R 数据挖掘蓝图》（Packt 出版社，2016 年）、《R：挖掘空间、文本、网络和社交媒体数据》（Packt 出版社，2017 年）、《PyTorch 秘籍》（Apress，2019 年）和《Python 实用说明性人工智能》（Apress，2022 年）。基于这些学术成果，他还在在线教育平台 Udemy 上开设了两门相关课程，为广大学员提供了宝贵的学习资源。

在 2018 年全球数据科学大会上，Pradeepta 发表了关于双向 LSTM 在时间序列预测中的主题演讲，为参会者提供了深入的技术见解。此外，他在 TED 演讲中探讨了人工智能对行业转型和变革工作角色的影响，引发了广泛的思考和讨论。Pradeepta 不仅在专业领域有所建树，还热衷于分享他的知识和经验。曾在各类聚会、技术机构、大学和社区论坛上进行了 150 余场技术演讲，为听众提供了宝贵的洞见。如需了解更多关于 Pradeepta 的信息，请访问他的 LinkedIn 页面（www.linkedin.com/in/pradeepta/）或关注他的 Twitter 账号 @pradmishra1。

关于技术审查员

Chris Thomas 是一位在人工智能和机器学习领域拥有丰富研究和开发经验的英国顾问。他还是英国分析师和程序员协会的专业会员，拥有超过 20 年的技术专业生涯经验。在此期间，他曾在公共部门、半导体行业、金融、公用事业和营销领域担任要职，积累了丰富的实践经验。

致　谢

我诚挚地感激我的妻子 Prajna，她长期以来的启迪和支持，甚至不惜牺牲自己的周末时间陪伴在我身边，为我完成这本书提供了巨大的帮助。同时，我也要感谢我的女儿 Aarya 和 Aadya，她们在我的整个写作过程中展现出无尽的包容与耐心，时刻支持着我。

此外，我还要特别感谢 Celestin Suresh John 和 Mark Powers。他们在整个出版过程中给予了我宝贵的帮助和指导，使我的方向得以修正，并加快了整个进程。

前 言

随着人工智能产品和解决方案的日益普及,基于图论的计算框架的需求也随之增长。为了使深度学习模型在实际应用中发挥更大的作用,我们需要一个动态的、灵活的,并且可与其他框架互操作的建模框架。这种需求促使我们寻求更加高效和可靠的计算框架,以支持人工智能技术的广泛应用。

PyTorch 是一款新兴的图计算工具/编程语言,旨在突破传统框架的限制。为了提升用户在深度学习模型部署方面的体验,以及在多模型(包括卷积神经网络、循环神经网络、LSTM以及深度神经网络)组合创建高级模型方面的便利性,PyTorch 提供了卓越的支持。

PyTorch 是由 Facebook 人工智能研究部门(FAIR)精心打造的,其设计理念在于使模型开发过程更为简洁、直观和动态。通过这种方式,开发者不再需要在编译和执行模型之前明确地声明各种对象,从而极大地提升了开发效率。PyTorch 是基于 Torch 框架的,并且是该框架的 Python 语言扩展版本。

本书适用于数据科学家、自然语言处理工程师、人工智能解决方案开发者、从事图计算框架工作的专业人士以及图论研究人员。本书旨在为读者提供张量基础知识及计算方面的入门指导。通过学习本书,读者将掌握如何运用 PyTorch 框架进行算术操作、矩阵代数和统计分布运算。

第 3 章和第 4 章详细描述了神经网络的基础知识。探讨了卷积神经网络、循环神经网络和 LSTM 等高级神经网络。读者将能够使用 PyTorch 函数实现这些模型。

第 5 章和第 6 章讨论了模型的微调、超参数调整以及生产环境中现有 PyTorch 模型的改进。读者将学习如何选择超参数来微调模型。

第 7 章讲解了自然语言处理。深度学习模型及其在自然语言处理和人工智能中的应用是业内要求最高的技能组合之一。读者将能够对执行和处理自然语言的 PyTorch 实现进行基准测试。此外,读者能够将 PyTorch 与其他基于图计算的深度学习编程工具进行比较。

源代码

访问 github.com/apress/pytorch-recipes-2e 可获取作者引用的所有源代码和其他补充材料。

目　录

译者序

关于作者

关于技术审查员

致谢

前言

第 1 章　PyTorch 入门，张量与张量运算 ·· 1

　　什么是 PyTorch ·· 5

　　PyTorch 安装 ·· 5

　　秘籍 1-1　张量的使用 ··· 7

　　小结 ··· 36

第 2 章　使用 PyTorch 中的概率分布 ·· 37

　　秘籍 2-1　采样张量 ·· 38

　　秘籍 2-2　可变张量 ·· 43

　　秘籍 2-3　统计学基础 ·· 45

　　秘籍 2-4　梯度计算 ·· 51

　　秘籍 2-5　张量运算之一 ·· 54

　　秘籍 2-6　张量运算之二 ·· 55

秘籍 2-7　统计分布 58

小结 62

第3章　使用 PyTorch 中的卷积神经网络和循环神经网络 63

秘籍 3-1　设置损失函数 63

秘籍 3-2　估计损失函数的导数 67

秘籍 3-3　模型微调 73

秘籍 3-4　优化函数选择 75

秘籍 3-5　进一步优化函数 80

秘籍 3-6　实现卷积神经网络 84

秘籍 3-7　模型重载 92

秘籍 3-8　实现循环神经网络 96

秘籍 3-9　实现用于回归问题的循环神经网络 102

秘籍 3-10　使用 PyTorch 内置的循环神经网络函数 104

秘籍 3-11　使用自编码器（Autoencoder） 108

秘籍 3-12　使用自编码器实现结果微调 113

秘籍 3-13　约束模型过拟合 116

秘籍 3-14　可视化模型过拟合 119

秘籍 3-15　初始化权重中的丢弃率 123

秘籍 3-16　添加数学运算 125

秘籍 3-17　循环神经网络中的嵌入层 128

小结 130

第4章　PyTorch 中的神经网络简介 131

秘籍 4-1　激活函数的使用 131

秘籍 4-2　激活函数可视化 139

秘籍 4-3　基本的神经网络模型 142

秘籍 4-4　张量微分 ·· 146

小结 ··· 148

第 5 章　PyTorch 中的监督学习 ··149

秘籍 5-1　监督模型的数据准备 ··· 153

秘籍 5-2　前向和反向传播神经网络 ·· 155

秘籍 5-3　优化和梯度计算 ··· 158

秘籍 5-4　查看预测结果 ·· 160

秘籍 5-5　监督模型逻辑回归 ·· 164

小结 ··· 168

第 6 章　使用 PyTorch 对深度学习模型进行微调 ····································169

秘籍 6-1　构建顺序神经网络 ·· 170

秘籍 6-2　确定批量的大小 ··· 172

秘籍 6-3　确定学习率 ··· 175

秘籍 6-4　执行并行训练 ·· 180

小结 ··· 182

第 7 章　使用 PyTorch 进行自然语言处理 ··183

秘籍 7-1　词嵌入 ··· 185

秘籍 7-2　使用 PyTorch 创建 CBOW 模型 ··· 190

秘籍 7-3　LSTM 模型 ··· 194

小结 ··· 199

第 8 章　分布式 PyTorch 建模、模型优化和部署 ··································200

秘籍 8-1　分布式 Torch 架构 ·· 200

秘籍 8-2	Torch 分布式组件	202
秘籍 8-3	设置分布式 PyTorch	203
秘籍 8-4	加载数据到分布式 PyTorch	205
秘籍 8-5	PyTorch 中的模型量化	208
秘籍 8-6	量化观察器应用	211
秘籍 8-7	使用 MNIST 数据集应用量化技术	213
小结		225

第 9 章 图像和音频的数据增强、特征工程和提取 226

秘籍 9-1	音频处理中的频谱图	226
秘籍 9-2	安装 Torchaudio	229
秘籍 9-3	将音频文件加载到 PyTorch 中	231
秘籍 9-4	安装用于音频的 Librosa 库	232
秘籍 9-5	频谱图变换	234
秘籍 9-6	Griffin-Lim 变换	236
秘籍 9-7	使用滤波器组进行梅尔尺度转换	237
秘籍 9-8	Librosa 的梅尔尺度转换与 Torchaudio 版本对比	240
秘籍 9-9	使用 Librosa 和 Torchaudio 进行 MFCC 和 LFCC	243
秘籍 9-10	图像数据增强	247
小结		250

第 10 章 PyTorch 模型可解释性和 Skorch 251

秘籍 10-1	安装 Captum 库	252
秘籍 10-2	主要归因：深度学习模型的特征重要性	253
秘籍 10-3	深度学习模型中神经元的重要性	259
秘籍 10-4	安装 Skorch 库	260

秘籍 10-5　Skorch 组件在神经网络分类器中的应用 ··· 262

秘籍 10-6　Skorch 神经网络回归器 ··· 265

秘籍 10-7　Skorch 模型的保存和加载 ·· 269

秘籍 10-8　使用 Skorch 创建神经网络模型流水线 ·· 270

秘籍 10-9　使用 Skorch 进行神经网络模型的
　　　　　 轮次评分 ·· 272

秘籍 10-10 使用 Skorch 进行超参数的网格搜索 ·· 274

小结 ··· 276

第 1 章

PyTorch 入门，张量与张量运算

PyTorch 已经逐渐发展成为编写动态模型的大型框架，其在数据科学家和工程师中非常受欢迎，广泛应用于部署大规模深度学习应用。本书为该领域的专家提供了一种实用且高效的处理模式，以解决实际数据科学问题。从日常生活中的众多应用程序可以看出，智能的广泛应用已深入到产品的各个功能层面，为用户带来了更优质的服务和体验。

在这个人工智能日益盛行的时代，我们正朝着充满未知的未来迈进。然而，只有通过开发具有训练能力的系统，才能充分释放人工智能的潜力。当数据维度较低且规模适中时，传统的机器学习方法往往能够展现出良好的效果。然而，面对高维度、大规模的数据，我们需要寻求新的解决方案。深度学习作为人工智能领域的一颗璀璨明星，正逐渐崭露头角。它以其独特的优势，能够处理海量复杂数据，为我们提供了更多的可能性。

PyTorch 是专为图形处理器（Graphics Processing Units，GPU）和（Central Processing Units，CPU）上的深度学习计算优化而设计的高性能张量库，旨在提升大规模计算环境中算法的性能。PyTorch 是基于 Python 和 Torch 工具的库，其中 Torch 是由 Facebook 人工智能研究部门开发的科学计算库。

在 GPU 上，基于 NumPy 的操作效率往往无法满足高计算负荷的需求，而静态深度学习库在计算灵活性和速度方面存在瓶颈。从实践者的角度看，PyTorch 中的张量与基于 Python 的 NumPy 库中的 N 维数组具有很高的相似性。为了提高

PyTorch 库在不同计算环境下的灵活性，我们提供了 NumPy 数组与张量数组之间相互转换的桥接选项。

PyTorch 广泛应用于各种使用场景，包括但不限于表格数据分析、自然语言处理、图像处理、计算机视觉、社交媒体数据分析和传感器数据处理。PyTorch 提供了大量的计算库和模块以支持这些应用，其中三个非常突出的模块分别是：

- **Autograd** 模块：该模块具备张量自动微分功能。在程序中，一个记录器类会记录操作，并通过 backward 触发检索这些操作以计算梯度。在实现神经网络模型时，该模块非常有用。

- **Optim** 模块：该模块提供了优化技术，用于最小化特定模型的误差函数。目前，PyTorch 支持多种高级优化方法，包括 Adam、随机梯度下降（SGD）等。

- **NN** 模块：NN（Neural Network）代表神经网络模型。在处理复杂的张量操作时，完全依赖手动定义的函数、层以及进一步的计算可能会带来一定的困扰，不仅难以记住，而且执行起来也可能比较繁琐。因此，PyTorch 库通过自动化生成一些函数，包括层、激活函数、损失函数和优化函数，从而极大地简化了这一过程。同时，还向用户提供了可以自定义的层，进一步减少了手动干预。而 NN 模块则内置了一套函数，能够自动化运行张量操作的手动过程，使得整个过程更为流畅和高效。

人工智能已广泛应用于各个行业，包括银行、金融服务、保险、医疗保健、制造业、零售业、临床试验和药物测试等。通过人工智能技术，可以实现物体分类、识别物体等功能，甚至可以检测欺诈行为。每个学习系统都包含输入数据、处理和输出层三个要素，这些要素之间相互关联，共同构成了人工智能的基础。随着训练时间的推移，通过机器学习系统不断学习新的示例或数据，系统的性能可以得到不断提升。然而，当系统变得过于复杂而难以反映现实数据时，则需要借助深度学习系统的力量。深度学习系统能够更好地处理复杂的任务，为我们提供更准确、更实用的解决方案。

在深度学习系统中，学习算法通常需要构建一个或多个隐藏层。在机器学习的领域内，主要涵盖了四种学习系统：监督学习、无监督学习、半监督学习和强化学习。其中，**监督学习**算法依赖于带有类别或结果标签的训练数据。我们向机器展示输入数据以及相应的标签或标记，机器通过一个函数来识别输入数据与标签之间的关系，该函数将输入与标签或标记进行关联。需要注意的是，这个函数的主要作用是建立输入与标签之间的映射关系。

在**无监督学习**过程中，我们仅将输入数据呈现给机器，并要求其根据相关性、相似性或差异性进行分组。无监督学习过程不需要预先标记的数据集，而是让机器自主学习数据的内在结构和模式。

在**半监督学习**过程中，我们向机器提供输入特征和带标签的数据或标记，并要求机器对未标记的结果或标签进行预测。

在**强化学习**过程中，我们采用奖励和惩罚机制来不断更新策略。在每一轮迭代中，对于正确的行为，通常会给予一定的奖励，而对于错误的行为，则会施加一定的惩罚。这种机制旨在维持策略的状态，并促进其不断向更好的方向发展。

在上述的机器学习算法示例中，我们均以数据集规模较小为前提。这是因为在实践中，获取大量标记过的数据是一项具有挑战性的任务。同时，传统的机器学习算法在进行大规模矩阵计算时往往需要较长的时间。鉴于机器学习算法在处理海量数据集时的限制，深度学习算法成了应对这一挑战的有效解决方案。

图 1-1 展示了机器学习、深度学习和人工智能之间的关系。自然语言处理是人工智能的重要组成部分，要求我们开发出能够理解自然语言并能够向代理提供响应的系统。以机器翻译为例，我们需要构建可以将一种语言（例如法语）的句子翻译成另一种语言（例如英语）的系统，反之亦然。为了实现这一目标，我们需要大量的英、法双语句子作为语料库。由于模型需要涵盖各种语言的细微差别，因此语料库的需求量非常大。

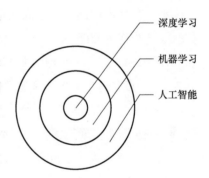

图 1-1　机器学习、深度学习和人工智能之间的关系

经过预先处理和特征构建后，为了获得最终的输出结果，需要进行大量的特征计算。在训练一个**机器学习**监督模型的过程中，往往需要数月的时间来完成并生成输出结果。为了实现可扩展性，**深度学习**算法如循环神经网络等被广泛应用。因此，人工智能与深度学习和机器学习之间存在着密切的联系。

在深度学习模型的实际应用中，我们面临着多重挑战。首先，深度学习高度依赖于大量的标记数据，数据质量对模型性能影响重大。其次，模型训练和推理需要高性能计算设备，以确保算法运行的效率和准确性。此外，智能算法的成熟度也是关键因素，当前大多数算法已较为完善。因此，要实现深度学习系统的成功应用，必须综合考虑高质量标记数据、先进计算设备和成熟智能算法等多方面因素。

以下是应用深度学习技术成功实现的各种案例：
- 语音识别：深度学习技术可以有效识别并转录语音，实现高精度的语音转文字。
- 视频分析：深度学习可以分析视频内容，提取关键信息，实现视频摘要、标签等功能。
- 视频中的异常检测：通过深度学习技术，可以检测视频中的异常事件，如火灾、交通事故等，提高监控系统的预警能力。
- 自然语言处理：深度学习技术可以实现文本分类、情感分析、机器翻译等功能，提高自然语言处理应用的性能。
- 机器翻译：深度学习技术可以高效地实现不同语言之间的翻译，提高翻译准确度和效率。
- 语音文本转换：通过深度学习技术，可以将语音转化为文字，实现语音笔记、语音写作等功能。

NVIDIA GPU 在处理大规模数据方面的应用，是一项具有里程碑意义的创新。为了实现在 GPU 环境下的编程，我们需引入特定的编程框架。目前，TensorFlow 与 PyTorch 是图形计算领域的两大主流框架。本书将深入探讨如何利用 PyTorch 框架实现数据科学算法与推理过程。

在图计算领域，PyTorch、TensorFlow 和 MXNet 是主流的框架。在神经计算领域的竞争中，PyTorch 和 TensorFlow 各有千秋。虽然两者在性能上大体相当，但在针对特定任务进行基准测试时，才能发现它们之间的差异性。此外，从概念角度看，这两个框架也有一些细微的差别，使得它们在不同应用场景中各有优势。

- 在 TensorFlow 中，明确定义张量、初始化会话以及为张量对象保留占位符是必要的操作。相对而言，在 PyTorch 中，这些步骤并非必需。
- 以情感分析为例，TensorFlow 和 PyTorch 在处理输入数据时的方法存在显著差异。在 TensorFlow 框架中，输入的句子需要预先标记为正例或负例。考虑到不同句子长度的差异，为了确保循环神经网络能够正常运作，开发者需要设定一个最大句子长度，并对其他长度不足的句子进行零填充。然而，在 PyTorch 中，这种处理方式被视为一种内置功能，因此用户无需自行设定句子的长度。

- 在 PyTorch 中，调试程序的过程相对简单且方便，而 TensorFlow 在这方面则显得较为复杂。
- 在数据可视化和模型部署方面，TensorFlow 展现出了显著的优势。然而，值得注意的是，PyTorch 正在持续取得进步，预计未来将具备与 TensorFlow 相当的功能。

经过多个版本的迭代更新，TensorFlow 现已成为稳定可靠的深度学习框架。相较之下，PyTorch 尚需进一步发展，才能实现框架的稳定性。尽管如此，PyTorch 在基于 Transformer 的大规模模型领域已成为行业标准，这些模型在 Hugging Face 平台上一应俱全。这些模型的应用范围广泛，可广泛应用于自然语言处理和计算机视觉任务。因此，对于寻找稳定且强大的深度学习框架的用户来说，PyTorch 无疑是一个值得考虑的选择。

什么是 PyTorch

PyTorch 是由 Facebook 人工智能研究部门（Facebook Artificial Intelligence Research，FAIR）开发的，专门用于机器学习和深度学习的工具。它广泛应用于处理大规模图像分析，包括目标检测、分割和分类等任务。然而，其应用领域并不仅限于此，还可与其他框架结合，实现更为复杂的算法。PyTorch 主要以 Python 和 C++ 编写，若在 GPU 环境下进行大规模计算，则需使用相应的编程语言进行修改。PyTorch 为编写在 GPU 环境中自动运行的函数，提供了一个极好的框架。

PyTorch 安装

在 Windows、Linux 或 macOS 中安装 PyTorch 非常简单方便，只需熟悉使用 Anaconda 和 Conda 环境管理软件包。以下步骤将描述如何在 Windows 环境中安装 PyTorch。

1. 打开 Anaconda Navigator 并转到环境（Environments）页面，如图 1-2 所示。

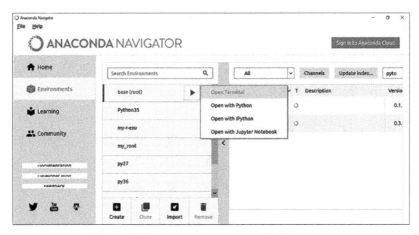

图 1-2　使用 Anaconda Navigator 安装 PyTorch

2. 打开终端并输入以下内容：

conda install -c peterjc123 pytorch

3. 启动 Jupyter 并打开 IPython Notebook[⊖]。

4. 输入以下命令检查 PyTorch 是否已安装：

In[1]:

from __future__ import print_function
import torch

5. 检查 PyTorch 版本。

In[2]:

torch.version.__version__

Out[2]:

1.12.1+cu113

以上安装过程是在 Microsoft Windows 操作系统上完成的。由于不同操作系统可能存在差异，请使用以下 URL 解决任何有关安装和错误的问题。

有两种安装 PyTorch 的方式：使用 Conda（Anaconda）库管理或者使用 Pip3 包管理框架。此外，本地系统（如 macOS、Windows 或 Linux）和云服务器（如 Microsoft Azure、AWS 和 GCP）的安装方法也是不同的。为了根据您的平台进行设置，请按照官方 PyTorch 安装文档上的指导进行操作，网址为 https://PyTorch.org/get-started/cloudpartners/。

⊖ 此处需要单击 "new"，新建一个 Python 3 Notebook 脚本文件。

PyTorch 包含各种组件：
- **Torch** 模块具有类似于 NumPy 的功能，并支持 GPU 运算。
- **Autograd** 模块中的 torch.autograd 提供了用于实现任意标量值函数的自动微分的类、方法和函数。只需要对现有代码进行最小的更改即可实现。并且只需要声明带有 requires_grad=True 关键字的 Tensor 类，就可以计算其梯度。
- **NN** 模块是 PyTorch 中的神经网络库。
- **Optim** 模块提供用于函数最小化和最大化的优化算法。
- **Multiprocessing** 模块是一个很有用的库，可用于在多个张量之间共享内存。
- **Utils** 模块具有加载数据的实用函数，以及其他功能。

接下来将继续本章内容的学习。

秘籍 1-1　张量的使用

问题

在 PyTorch 中，数据结构主要基于图和张量，因此，掌握张量的定义以及基本操作如索引、改变形状和更改数据类型等，对于理解和应用 PyTorch 至关重要。

解决方案

为了解决这个问题，应深入练习张量及其操作，并掌握各种操作的实例。尽管我们假设读者已具备 PyTorch 和 Python 的基础知识，但回顾 PyTorch 的相关知识对于激发新手的兴趣仍然至关重要。

编程实战

接下来，将通过一些基础示例来展示张量和张量操作，其中涵盖了数学运算的相关内容。

对象 x 是一个列表（list）。可以使用以下语法检查 Python 对象是否为张量对象。通常，**is_tensor** 函数检查对象是否为张量对象，而 **is_storage** 函数检查对象是否存储为张量对象。

In[3]:

x = [12,23,34,45,56,67,78]

In[4]:

定义一个标量（scalar）

scalar = torch.tensor(10)

scalar

Out[4]:

tensor(10)

In[5]:

scalar.ndim

Out[5]:

0

In[6]:

scalar.item()

Out[6]:

10

In[7]:

定义一个向量（vector）

vector = torch.tensor([5,5])

vector

Out[7]:

tensor([5, 5])

In[8]:

vector.ndim

Out[8]:

1

In[9]:

vector.shape

Out[9]:

torch.Size([2])

In[10]:

# 定义一个矩阵（matrix）
matrix = torch.tensor([[4, 5],
                       [10, 110]])
matrix

Out[10]:

tensor([[4, 5], [10, 110]])

In[11]:

matrix.ndim

Out[11]:

2

In[12]:

matrix.shape

Out[12]:

torch.Size([2, 2])

In[13]:

# 定义一个张量（tensor），张量是多元的
tensor = torch.tensor([[[4,64, 5,4],
                        [10,20,30, 110],
                        [45,34,67,40],
                        [56,67,89,90]]])
tensor

Out[13]:

tensor([[[4, 64, 5, 4], [10, 20, 30, 110], [45, 34, 67, 40], [56, 67, 89, 90]]])

In[14]:

tensor.ndim

Out[14]:

3

In[15]:

tensor.shape

Out[15]:
torch.Size([1, 4, 4])

In[16]:
tensor.dtype

Out[16]:
torch.int64

In[17]:
tensor.device

Out[17]:
device(type='cpu')

In[18]:
torch.is_tensor(x)

Out[18]:
False

In[19]:
torch.is_storage(x)

Out[19]:
False

接下来，将参照 NumPy 库的方法，使用 Torch 生成一个随机数对象。在此过程中，同样需要对新创建的对象进行张量以及存储类型的检查。

In[20]:
y = torch.randn(1,2,3,4,5)

In[21]:
torch.is_tensor(y)

Out[21]:
True

In[22]:
torch.is_storage(y)

Out[22]:
False

In[23]:

torch.numel(y)　# 输入张量 y 中的元素总数

Out[23]:

120

从上述代码中,可以明确地了解到,对象 y 代表了一个张量,但是该张量并没有被实际存储。为了确定该张量中包含的元素总数,建议使用 numel 函数(用于计算数值元素)。以下是一个示例脚本,用于创建一个二维全零张量并计算其中的数值元素数量。

In[24]:

torch.zeros(4,4)

Out[24]:

tensor([[0., 0., 0., 0.], [0., 0., 0., 0.], [0., 0., 0., 0.], [0., 0., 0., 0.]])

In[25]:

torch.numel(torch.zeros(4,4))

Out[25]:

16

In[26]:

torch.eye(3)

Out[26]:

tensor([[1., 0., 0.], [0., 1., 0.], [0., 0., 1.]])

In[27]:

torch.eye(5)

Out[27]:

tensor([[1., 0., 0., 0., 0.], [0., 1., 0., 0., 0.], [0., 0., 1., 0., 0.], [0., 0., 0., 1., 0.], [0., 0., 0., 0., 1.]])

类似于 NumPy 中的操作,eye 函数能生成一个对角矩阵。此矩阵的特点是其对角线上的元素为 1,而非对角线上的元素为 0。在调用 eye 函数时,需要提供相应的矩阵形状作为参数。以下是一个关于如何提供矩阵形状参数的示例。

In[28]:

torch.eye(3,4)

Out[28]:

tensor([[1., 0., 0., 0.], [0., 1., 0., 0.], [0., 0., 1., 0.]])

In[29]:

torch.eye(5,4)

Out[29]:

tensor([[1., 0., 0., 0.], [0., 1., 0., 0.], [0., 0., 1., 0.], [0., 0., 0., 1.], [0., 0., 0., 0.]])

In[30]:

type(x)

Out[30]:

list

在处理线性空间和其中的点时，可以通过张量操作进行创建。以一个具体示例为例，可以创建一个包含从2到10范围内25个点的线性空间。值得注意的是，Torch能够直接处理NumPy数组格式的数据。

In[31]:

import numpy as np
x1 = np.array(x)

In[32]:

x1

Out[32]:

array([12, 23, 34, 45, 56, 67, 78])

In[33]:

torch.from_numpy(x1)

Out[33]:

tensor([12, 23, 34, 45, 56, 67, 78])

In[34]:

torch.linspace(2, 10, steps=25)   # 线性间距

Out[34]:

tensor([2.0000, 2.3333, 2.6667, 3.0000, 3.3333, 3.6667, 4.0000, 4.3333, 4.6667, 5.0000, 5.3333, 5.6667, 6.0000, 6.3333, 6.6667, 7.0000, 7.3333, 7.6667, 8.0000, 8.3333, 8.6667, 9.0000, 9.3333, 9.6667, 10.0000])

In[35]:

torch.linspace(-10, 10, steps=15)

Out[35]:
tensor([-1.0000e+01, -8.5714e+00, -7.1429e+00, -5.7143e+00, -4.2857e+00,
-2.8571e+00, -1.4286e+00, -2.3842e-07, 1.4286e+00, 2.8571e+00, 4.2857e+00,
5.7143e+00, 7.1429e+00, 8.5714e+00, 1.0000e+01])

与线性间距类似，可以创建对数间距。

In[36]:

torch.logspace(start=-10, end=10, steps=15) # 对数间距

Out[36]:

tensor([1.0000e-10, 2.6827e-09, 7.1969e-08, 1.9307e-06, 5.1795e-05,
1.3895e-03, 3.7276e-02, 1.0000e+00, 2.6827e+01, 7.1969e+02, 1.9307e+04,
5.1795e+05, 1.3895e+07, 3.7276e+08, 1.0000e+10])

以下是一维全1张量和二维全1张量的创建示例。

In[37]:

torch.ones(4)

Out[37]:

tensor([1., 1., 1., 1.])

In[38]:

torch.ones(4,5)

Out[38]:

tensor([[1., 1., 1., 1., 1.], [1., 1., 1., 1., 1.], [1., 1., 1., 1., 1.],
[1., 1., 1., 1., 1.]])

在数据科学领域，随机数生成是一个常见的操作，用于创建或收集空间中的样本数据点，以便模拟数据的内在结构。这些随机数可以从统计分布、任意两个数值或预定义的分布中获取。以 PyTorch 为例，可以使用类似 NumPy 的函数来生成随机数。**均匀分布**特指每个结果出现概率相等的分布，这意味着事件的概率是恒定的。

In[39]:

#0和1之间均匀分布的随机数

torch.rand(10)

Out[39]:

tensor([0.1408, 0.4445, 0.4251, 0.2663, 0.3743, 0.4784, 0.3760, 0.1876,
0.2151, 0.6876])

以下是脚本示例,演示了如何生成 0 和 1 之间的随机数。生成的结果张量可以重新塑形为(4,5)矩阵。同时,还可以生成来自正态分布的随机数,该分布的算术平均值为 0,标准差为 1,如下所示。

使用随机置换从值范围中选择随机值需要先定义范围。这个范围可以通过使用 torch.arange 函数来创建。使用 torch.arange 函数时,必须定义步长,将所有值放置在等距空间中。默认情况下,步长为 1。

In[40]:

torch.rand(4, 5)

生成一个大小为4行5列的矩阵,并用0和1之间的随机值填充

Out[40]:

tensor([[0.2733, 0.0302, 0.8835, 0.9537, 0.9662], [0.6296, 0.3106, 0.4029, 0.8133, 0.1697], [0.8578, 0.6517, 0.0440, 0.6197, 0.9889], [0.8614, 0.6288, 0.2158, 0.4593, 0.2444]])

In[41]:

从正态分布中随机抽取数字
平均值为 0,标准差为 1

torch.randn(10)

Out[41]:

tensor([1.0115, -0.7502, 1.1994, 0.8736, 0.5633, -0.7702, 0.1826, -1.9931, 0.5159, 0.1521])

In[42]:

torch.randn(4, 5)

Out[42]:

tensor([[0.3744, 2.1839, -1.8229, 1.0682, 1.5394], [0.9689, -1.3085, -0.3300, 0.3960, -0.6079], [2.3485, 1.2880, 0.6754, -2.0426, -0.3121], [-0.4897, -1.5335, 0.0467, -0.6213, 1.7185]])

In[43]:

从一定范围内选择值,这被称为随机排列

torch.randperm(10)

Out[43]:

tensor([1, 6, 3, 2, 0, 8, 4, 5, 7, 9])

In[44]:

range 函数的用法
torch.arange(10, 40,2)　# 步长为 2

Out[44]:

tensor([10, 12, 14, 16, 18, 20, 22, 24, 26, 28, 30, 32, 34, 36, 38])

In[45]:

torch.arange(10,40)　# 步长为 1

Out[45]:

tensor([10, 11, 12, 13, 14, 15, 16, 17, 18, 19, 20, 21, 22, 23, 24, 25, 26, 27, 28, 29, 30, 31, 32, 33, 34, 35, 36, 37, 38, 39])

要查找一维张量中的最小值和最大值，可以使用 argmin 和 argmax。如果输入是矩阵，则需要指定搜索的维度，以便在行或列中搜索最小（或最大）值。

In[46]:

d = torch.randn(4, 5)
d

Out[46]:

tensor([[1.0085, -0.8545, -0.6958, 1.6716, -0.0118], [0.2134, 1.1154, -0.6426, -1.3651, -1.5724], [0.2452, 0.8356, 2.0297, -0.2397, 0.8560], [0.9786, -0.8538, -0.6449, 0.3903, 1.5966]])

In[47]:

torch.argmin(d,dim=1)

Out[47]:

tensor([1, 4, 3, 1])

In[48]:

torch.argmax(d,dim=1)

Out[48]:

tensor([3, 1, 2, 4])

当输入数据仅为一行或一列时，它被视为一个单独的维度，称为一维张量或向量。当输入数据包含多行和多列形成一个矩阵时，它被称为二维张量。对于超过二维的维度，我们称之为多维张量。

In[49]:

创建一个填充值为 0 的二维张量

torch.zeros(4,5)

Out[49]:
tensor([[0., 0., 0., 0., 0.], [0., 0., 0., 0., 0.], [0., 0., 0., 0., 0.], [0., 0., 0., 0., 0.]])

In[50]:
# 创建一个填充值为 0 的一维张量
torch.zeros(10)

Out[50]:
tensor([0., 0., 0., 0., 0., 0., 0., 0., 0., 0.])

接下来，将演示如何生成一个二维张量的示例，并通过 concat 操作进行索引和连接。

In[51]:
# 对张量进行索引和执行操作
x = torch.randn(4,5)

In[52]:
x

Out[52]:
tensor([[-1.5343, -1.3533, -0.8621, -1.1674, -0.1114], [0.2790, 0.0463, 1.5364, -0.1287, 0.6379], [-0.4542, 0.5196, 0.2335, -0.5135, -0.6602], [-0.6930, 0.0541, -0.8463, -0.4498, -0.0280]])

In[53]:
# 连接两个张量
torch.cat((x,x))

Out[53]:
tensor([[-1.5343, -1.3533, -0.8621, -1.1674, -0.1114], [0.2790, 0.0463, 1.5364, -0.1287, 0.6379], [-0.4542, 0.5196, 0.2335, -0.5135, -0.6602], [-0.6930, 0.0541, -0.8463, -0.4498, -0.0280], [-1.5343, -1.3533, -0.8621, -1.1674, -0.1114], [0.2790, 0.0463, 1.5364, -0.1287, 0.6379], [-0.4542, 0.5196, 0.2335, -0.5135, -0.6602], [-0.6930, 0.0541, -0.8463, -0.4498, -0.0280]])

样本 x 张量也可以用于三维情况。同样地，有两种不同的方法可以创建三维张量；第三维可以沿行或列进行扩展。

In[54]:
根据数组大小连接 n 次

```
torch.cat((x,x,x))
```

Out[54]:

tensor([[-1.5343, -1.3533, -0.8621, -1.1674, -0.1114], [0.2790, 0.0463, 1.5364, -0.1287, 0.6379], [-0.4542, 0.5196, 0.2335, -0.5135, -0.6602], [-0.6930, 0.0541, -0.8463, -0.4498, -0.0280], [-1.5343, -1.3533, -0.8621, -1.1674, -0.1114], [0.2790, 0.0463, 1.5364, -0.1287, 0.6379], [-0.4542, 0.5196, 0.2335, -0.5135, -0.6602], [-0.6930, 0.0541, -0.8463, -0.4498, -0.0280], [-1.5343, -1.3533, -0.8621, -1.1674, -0.1114], [0.2790, 0.0463, 1.5364, -0.1287, 0.6379], [-0.4542, 0.5196, 0.2335, -0.5135, -0.6602], [-0.6930, 0.0541, -0.8463, -0.4498, -0.0280]])

In[55]:

```
# 根据数组大小，按列连接 n 次
torch.cat((x,x,x),1)
```

Out[55]:

tensor([[-1.5343, -1.3533, -0.8621, -1.1674, -0.1114, -1.5343, -1.3533, -0.8621, -1.1674, -0.1114, -1.5343, -1.3533, -0.8621, -1.1674, -0.1114], [0.2790, 0.0463, 1.5364, -0.1287, 0.6379, 0.2790, 0.0463, 1.5364, -0.1287, 0.6379, 0.2790, 0.0463, 1.5364, -0.1287, 0.6379], [-0.4542, 0.5196, 0.2335, -0.5135, -0.6602, -0.4542, 0.5196, 0.2335, -0.5135, -0.6602, -0.4542, 0.5196, 0.2335, -0.5135, -0.6602], [-0.6930, 0.0541, -0.8463, -0.4498, -0.0280, -0.6930, 0.0541, -0.8463, -0.4498, -0.0280, -0.6930, 0.0541, -0.8463, -0.4498, -0.0280]])

In[56]:

```
# 根据数组大小，按行连接 n 次
torch.cat((x,x),0)
```

Out[56]:

tensor([[-1.5343, -1.3533, -0.8621, -1.1674, -0.1114], [0.2790, 0.0463, 1.5364, -0.1287, 0.6379], [-0.4542, 0.5196, 0.2335, -0.5135, -0.6602], [-0.6930, 0.0541, -0.8463, -0.4498, -0.0280], [-1.5343, -1.3533, -0.8621, -1.1674, -0.1114], [0.2790, 0.0463, 1.5364, -0.1287, 0.6379], [-0.4542, 0.5196, 0.2335, -0.5135, -0.6602], [-0.6930, 0.0541, -0.8463, -0.4498, -0.0280]])

In[57]:

```
# 如何将张量分割成小块
```

```
torch.arange(11).chunk(6)
    (tensor([0, 1]),
     tensor([2, 3]),
     tensor([4, 5]),
     tensor([6, 7]),
     tensor([8, 9]),
     tensor([10]))

torch.arange(12).chunk(6)
    (tensor([0, 1]),
     tensor([2, 3]),
     tensor([4, 5]),
     tensor([6, 7]),
     tensor([8, 9]),
     tensor([10, 11]))

torch.arange(13).chunk(6)
    (tensor([0, 1, 2]),
     tensor([3, 4, 5]),
     tensor([6, 7, 8]),
     tensor([9, 10, 11]),
     tensor([12]))
```

在处理张量时，可以根据需要将其分割成若干个块。这些块可以在行或列的维度上进行创建。例如，对于一个大小为（4，4）的张量样本，通过调整 chunk 函数的第三个参数为 0 或 1，可以灵活地创建所需的块。

In [58]:
```
a = torch.randn(4, 4)
print(a)
```

```
torch.chunk(a,2)
tensor([[-0.5899, -1.3432, -1.0576, -0.1696],
        [ 0.2623, -0.1585,  1.0178, -0.2216],
        [-1.1716, -1.2771,  0.8073, -0.7717],
        [ 0.1768,  0.6423, -0.3200, -0.0480]])
```

Out[58]:
```
(tensor([[-0.5899, -1.3432, -1.0576, -0.1696],
         [ 0.2623, -0.1585,  1.0178, -0.2216]]),
 tensor([[-1.1716, -1.2771,  0.8073, -0.7717],
         [ 0.1768,  0.6423, -0.3200, -0.0480]]))
```

In[59]:

torch.chunk(a,2,0)

Out[59]:

(tensor([[-0.5899, -1.3432, -1.0576, -0.1696], [0.2623, -0.1585, 1.0178, -0.2216]]), tensor([[-1.1716, -1.2771, 0.8073, -0.7717], [0.1768, 0.6423, -0.3200, -0.0480]]))

In[60]:

torch.chunk(a,2,1)

Out[60]:

(tensor([[-0.5899, -1.3432], [0.2623, -0.1585], [-1.1716, -1.2771], [0.1768, 0.6423]]), tensor([[-1.0576, -0.1696], [1.0178, -0.2216], [0.8073, -0.7717], [-0.3200, -0.0480]]))

In[61]:

torch.Tensor([[11,12],[23,24]])

Out[61]:

tensor([[11., 12.], [23., 24.]])

在 PyTorch 中，gather 函数通过使用索引参数，从给定的张量中提取相应位置的元素，并将这些元素整合到一个新的张量中。索引位置的确定需要使用 PyTorch 中的 LongTensor 函数。

In[62]:

torch.gather(torch.Tensor([[11,12],[23,24]]), 1,
            torch.LongTensor([[0,0],[1,0]]))

Out[62]:

tensor([[11., 11.], [24., 23.]])

In[63]:

torch.LongTensor([[0,0],[1,0]])

包含索引的一维张量

Out[63]:

tensor([[0, 0], [1, 0]])

在处理张量时，LongTensor 函数或 index_select 函数可以用来从特定张量中提取相关的值。通过以下示例代码，将展示两种不同的选择方式：按行选择和按列选择。如果第二个参数设置为 0，则表示按行进行选择；如果设置为 1，则表示按列进行选择。

In [64]:
```
a = torch.randn(4, 4)
print(a)
tensor([[-0.9183, -2.3470,  1.5208, -0.1585],
        [-0.6741, -0.6297,  0.2581, -1.1954],
        [ 1.0443, -1.3408,  0.7863, -0.6056],
        [-0.6946, -0.5963,  0.1936, -2.0625]])
```

In [65]:
```
indices = torch.LongTensor([0, 2])
```

In [66]:
```
torch.index_select(a, 0, indices)
```

Out [66]:
```
tensor([[-0.9183, -2.3470, 1.5208, -0.1585], [ 1.0443, -1.3408, 0.7863, -0.6056]])
```

In [67]:
```
torch.index_select(a, 1, indices)
```

Out [67]:
```
tensor([[-0.9183, 1.5208], [-0.6741, 0.2581], [ 1.0443, 0.7863], [-0.6946, 0.1936]])
```

在处理张量时，检查非缺失值是一种常见的操作。其主要目标是在大张量结构中识别非零元素。

In [68]:
```
# 使用nonzero函数识别输入张量中的非零元素
torch.nonzero(torch.tensor([10,00,23,0,0.0]))
```

Out [68]:
```
tensor([[0], [2]])
```

In [69]:
```
torch.nonzero(torch.Tensor([10,00,23,0,0.0]))
```

Out [69]:
```
tensor([[0], [2]])
```

将输入张量进行重构，使其变得更小，不仅会加速计算过程，而且对于分布式计算也有所帮助。在此过程中，使用 split 函数能将长张量拆分成多个更小的张量。

In[70]:

#将张量分割成小块，每两个元素分割

torch.split(torch.tensor([12,21,34,32,45,54,56,65]),2)

Out[70]:

(tensor([12, 21]), tensor([34, 32]), tensor([45, 54]), tensor([56, 65]))

In[71]:

#将张量分割成小块，每三个元素分割

torch.split(torch.tensor([12,21,34,32,45,54,56,65]),3)

Out[71]:

(tensor([12, 21, 34]), tensor([32, 45, 54]), tensor([56, 65]))

In[72]:

torch.zeros(3,2,4)

Out[72]:

tensor([[[0., 0., 0., 0.], [0., 0., 0., 0.]], [[0., 0., 0., 0.], [0., 0., 0., 0.]], [[0., 0., 0., 0.], [0., 0., 0., 0.]]])

In[73]:

torch.zeros(3,2,4).size()

Out[73]:

torch.Size([3, 2, 4])

在面对计算难度给定的情况下，为了更好地调整输入张量的大小，常常需要使用 transpose 函数来进行张量的重塑。值得注意的是，transpose 函数有两种表达方式：.t 和 .transpose。

In[74]:

#沿新的维度重塑张量

In[75]:

x

Out[75]:

tensor([[-1.5343, -1.3533, -0.8621, -1.1674, -0.1114], [0.2790, 0.0463, 1.5364, -0.1287, 0.6379], [-0.4542, 0.5196, 0.2335, -0.5135, -0.6602], [-0.6930, 0.0541, -0.8463, -0.4498, -0.0280]])

In[76]:
x.t() # transpose 是调整张量形状的一种方式

Out[76]:
tensor([[-1.5343, 0.2790, -0.4542, -0.6930], [-1.3533, 0.0463, 0.5196, 0.0541], [-0.8621, 1.5364, 0.2335, -0.8463], [-1.1674, -0.1287, -0.5135, -0.4498], [-0.1114, 0.6379, -0.6602, -0.0280]])

In[77]:
基于行和列的部分 transpose

In[78]:
x.transpose(1,0)

Out[78]:
tensor([[-1.5343, 0.2790, -0.4542, -0.6930], [-1.3533, 0.0463, 0.5196, 0.0541], [-0.8621, 1.5364, 0.2335, -0.8463], [-1.1674, -0.1287, -0.5135, -0.4498], [-0.1114, 0.6379, -0.6602, -0.0280]])

在 PyTorch 中，unbind 函数被用于从张量中移除一个维度。当需要移除行维度时，将参数 dim 的值设定为 0；当需要移除列维度时，则将参数 dim 的值设定为 1。

In[79]:
从张量中删除一个维度

In[80]:
x

Out[80]:
tensor([[-1.5343, -1.3533, -0.8621, -1.1674, -0.1114], [0.2790, 0.0463, 1.5364, -0.1287, 0.6379], [-0.4542, 0.5196, 0.2335, -0.5135, -0.6602], [-0.6930, 0.0541, -0.8463, -0.4498, -0.0280]])

In[81]:
torch.unbind(x,1) # dim=1 删除一列

Out[81]:
(tensor([-1.5343, 0.2790, -0.4542, -0.6930]), tensor([-1.3533, 0.0463, 0.5196, 0.0541]), tensor([-0.8621, 1.5364, 0.2335, -0.8463]), tensor([-1.1674, -0.1287, -0.5135, -0.4498]), tensor([-0.1114, 0.6379, -0.6602, -0.0280]))

In[82]:

torch.unbind(x)   #dim=0 删除一行

Out[82]:

(tensor([-1.5343, -1.3533, -0.8621, -1.1674, -0.1114]), tensor([0.2790, 0.0463, 1.5364, -0.1287, 0.6379]), tensor([-0.4542, 0.5196, 0.2335, -0.5135, -0.6602]), tensor([-0.6930, 0.0541, -0.8463, -0.4498, -0.0280]))

In[83]:

x

Out[83]:

tensor([[-1.5343, -1.3533, -0.8621, -1.1674, -0.1114], [0.2790, 0.0463, 1.5364, -0.1287, 0.6379], [-0.4542, 0.5196, 0.2335, -0.5135, -0.6602], [-0.6930, 0.0541, -0.8463, -0.4498, -0.0280]])

在 PyTorch 中，数学函数在算法实现中扮演着核心角色。接下来，我们将深入探讨一些有助于执行算术操作的函数。简单来说，标量就是一个单独的数值。而一维张量，可以看作是一个数字序列。使用 add 和 mul 函数，可以对一维张量进行标量加法和乘法运算。以下脚本将展示如何进行这些操作的示例。

In[84]:

计算基本数学函数

In[85]:

torch.abs(torch.FloatTensor([-10, -23, 3.000]))

Out[85]:

tensor([10., 23., 3.])

In[86]:

给现有张量增加值，标量加法

torch.add(x,20)

Out[86]:

tensor([[18.4657, 18.6467, 19.1379, 18.8326, 19.8886], [20.2790, 20.0463, 21.5364, 19.8713, 20.6379], [19.5458, 20.5196, 20.2335, 19.4865, 19.3398], [19.3070, 20.0541, 19.1537, 19.5502, 19.9720]])

In[87]:

x

Out[87]:

tensor([[-1.5343, -1.3533, -0.8621, -1.1674, -0.1114], [0.2790, 0.0463, 1.5364, -0.1287, 0.6379], [-0.4542, 0.5196, 0.2335, -0.5135, -0.6602], [-0.6930, 0.0541, -0.8463, -0.4498, -0.0280]])

In[88]:

标量乘法

torch.mul(x,2)

Out[88]:

tensor([[-3.0686, -2.7065, -1.7242, -2.3349, -0.2227], [0.5581, 0.0926, 3.0727, -0.2575, 1.2757], [-0.9084, 1.0392, 0.4670, -1.0270, -1.3203], [-1.3859, 0.1082, -1.6926, -0.8995, -0.0560]])

In[89]:

x

Out[89]:

tensor([[-1.5343, -1.3533, -0.8621, -1.1674, -0.1114], [0.2790, 0.0463, 1.5364, -0.1287, 0.6379], [-0.4542, 0.5196, 0.2335, -0.5135, -0.6602], [-0.6930, 0.0541, -0.8463, -0.4498, -0.0280]])

可以使用以下示例脚本来完成组合数学运算，例如将线性方程表达为张量运算。具体而言，将结果对象 y 表示为 beta 值乘以自变量 x 对象的线性组合，同时再加上一个常数项。

In[90]:

将方程表示为张量的形式

In[91]:

y = intercept + (beta * x)

In[92]:

intercept = torch.randn(1)
intercept

Out[92]:

tensor([-1.1444])

In[93]:

x = torch.randn(2, 2)
x

Out[93]:

tensor([[1.3517, -0.3991], [-0.4170, -0.1862]])

In[94]:

beta = 0.7456
beta

Out[94]:

0.7456

  Output = Constant + (beta * Independent)

In[95]:

torch.mul(x,beta)

Out[95]:

tensor([[1.0078, -0.2976], [-0.3109, -0.1388]])

In[96]:

torch.add(x,beta,intercept)

Out[96]:

tensor([[0.4984, -1.2524], [-1.2703, -1.0395]])

In[97]:

torch.mul(intercept,x)

Out[97]:

tensor([[-1.5469, 0.4568], [0.4773, 0.2131]])

In[98]:

torch.mul(x,beta)

Out[98]:

tensor([[1.0078, -0.2976], [-0.3109, -0.1388]])

In[99]:

y = intercept + (beta * x)
torch.add(torch.mul(intercept,x),torch.mul(x,beta))   # 张量 y

Out[99]:

tensor([[-0.5391, 0.1592], [0.1663, 0.0743]])

  类似于 Numpy 的操作，也可以使用张量进行逐元素矩阵乘法。这里有两种不同的矩阵乘法方式：逐元素乘法和点乘法。逐元素乘法是指对两个矩阵中的对应元素进行相乘,而点乘法则是对应位置的向量的点积运算。在实际使用中,应该根据实际需求选择适合的矩阵乘法方式。

In[100]:

tensor

Out[100]:

tensor([[[4, 64, 5, 4], [10, 20, 30, 110], [45, 34, 67, 40], [56, 67, 89, 90]]])

In[101]:

逐元素矩阵乘法
tensor * tensor

Out[101]:

tensor([[[16, 4096, 25, 16], [100, 400, 900, 12100], [2025, 1156, 4489, 1600], [3136, 4489, 7921, 8100]]])

In[102]:

torch.matmul(tensor, tensor)

Out[102]:

tensor([[[1105, 1974, 2631, 7616], [7750, 9430, 12450, 13340], [5775, 8518, 9294, 10200], [9939, 13980, 16263, 19254]]])

In[103]:

tensor @ tensor

Out[103]:

tensor([[[1105, 1974, 2631, 7616], [7750, 9430, 12450, 13340], [5775, 8518, 9294, 10200], [9939, 13980, 16263, 19254]]])

类似于 NumPy 操作,张量值必须使用以下语法进行向上或向下取整函数的舍入操作。

In[104]:

# 四舍五入张量值
torch.manual_seed(1234)
torch.randn(5,5)

Out[104]:

tensor([[-0.1117, -0.4966, 0.1631, -0.8817, 0.0539], [0.6684, -0.0597, -0.4675, -0.2153, -0.7141], [-1.0831, -0.5547, 0.9717, -0.5150, 1.4255], [0.7987, -1.4949, 1.4778, -0.1696, -0.9919], [-1.4569, 0.2563, -0.4030, 0.4195, 0.9380]])

In [105]:
torch.manual_seed(1234)
torch.ceil(torch.randn(5,5))

Out[105]:
tensor([[-0., -0., 1., -0., 1.], [1., -0., -0., -0., -0.], [-1., -0., 1., -0., 2.], [1., -1., 2., -0., -0.], [-1., 1., -0., 1., 1.]])

In [106]:
torch.manual_seed(1234)
torch.floor(torch.randn(5,5))

Out[106]:
tensor([[-1., -1., 0., -1., 0.], [0., -1., -1., -1., -1.], [-2., -1., 0., -1., 1.], [0., -2., 1., -1., -1.], [-2., 0., -1., 0., 0.]])

通过使用最小值和最大值参数以及 clamp 函数，可以将任何张量的值限制在一定范围内。无论是一维张量还是二维张量，同一函数可以同时设置最小值和最大值，或只设置其中的任何一个。一维张量中的实现相对简单。以下示例展示了在二维张量场景中的实现方式。

In [107]:
# 将数值截断在一个区间范围内，例如 [–0.3, 0.4]
torch.manual_seed(1234)
torch.clamp(torch.floor(torch.randn(5,5)), min=-0.3, max=0.4)

Out[107]:
tensor([[-0.3000, -0.3000, 0.0000, -0.3000, 0.0000], [0.0000, -0.3000, -0.3000, -0.3000, -0.3000], [-0.3000, -0.3000, 0.0000, -0.3000, 0.4000], [0.0000, -0.3000, 0.4000, -0.3000, -0.3000], [-0.3000, 0.0000, -0.3000, 0.0000, 0.0000]])

In [108]:
# 仅使用下限截断
torch.manual_seed(1234)
torch.clamp(torch.floor(torch.randn(5,5)), min=-0.3)

Out[108]:
tensor([[-0.3000, -0.3000, 0.0000, -0.3000, 0.0000], [0.0000, -0.3000, -0.3000, -0.3000, -0.3000], [-0.3000, -0.3000, 0.0000, -0.3000, 1.0000], [0.0000, -0.3000, 1.0000, -0.3000, -0.3000], [-0.3000, 0.0000, -0.3000, 0.0000, 0.0000]])

In [109]:
# 仅使用上限截断
torch.manual_seed(1234)
torch.clamp(torch.floor(torch.randn(5,5)), max=0.3)

Out[109]:
tensor([[-1.0000, -1.0000, 0.0000, -1.0000, 0.0000], [0.0000, -1.0000, -1.0000, -1.0000, -1.0000], [-2.0000, -1.0000, 0.0000, -1.0000, 0.3000], [0.0000, -2.0000, 0.3000, -1.0000, -1.0000], [-2.0000, 0.0000, -1.0000, 0.0000, 0.0000]])

张量的指数如何获取呢？如果张量包含小数位并被定义为浮点数据类型，则如何获取其分数部分？接下来的示例中将给出解答。

In [110]:
# 标量除法
torch.div(x,0.10)

Out[110]:
tensor([[13.5168, -3.9914], [-4.1705, -1.8621]])

In [111]:
# 计算张量的指数
torch.exp(x)

Out[111]:
tensor([[3.8639, 0.6709], [0.6590, 0.8301]])

In [112]:
np.exp(x)

Out[112]:
tensor([[3.8639, 0.6709], [0.6590, 0.8301]])

In [113]:
获取每个张量的分数部分

In [114]:
torch.add(x,10)

Out[114]:
tensor([[11.3517, 9.6009], [9.5830, 9.8138]])

In [115]:

torch.frac(torch.add(x,10))

Out[115]:

tensor([[0.3517, 0.6009], [0.5830, 0.8138]])

以下语法解释了张量中的对数值计算方式。带负号的值将被转换为 nan。power 函数可以计算张量中任何值的指数。

In [116]:

计算张量中值的对数

In [117]:

x

Out[117]:

tensor([[1.3517, -0.3991], [-0.4170, -0.1862]])

In [118]:

torch.log(x)   # 负数的对数是 nan

Out[118]:

tensor([[0.3013, nan], [nan, nan]])

In [119]:

针对负值问题,可使用幂函数转换予以修正

torch.pow(x,2)

Out[119]:

tensor([[1.8270, 0.1593], [0.1739, 0.0347]])

In [120]:

四舍五入类似于 numpy 中的运算

In [121]:

x

Out[121]:

tensor([[1.3517, -0.3991], [-0.4170, -0.1862]])

In [122]:

np.round(x)

Out[122]:

tensor([[1., -0.], [-0., -0.]])

In [123]:
torch.round(x)

Out[123]:
tensor([[1., -0.], [-0., -0.]])

要计算变换函数（即 Sigmoid、双曲正切和径向基函数，这些函数是深度学习中最常用的传递函数），则必须构建张量。以下示例脚本展示了如何创建一个 Sigmoid 函数，并将其应用于张量。

In [124]:
计算输入张量的 sigmoid 值

In [125]:
x

Out[125]:
tensor([[1.3517, -0.3991], [-0.4170, -0.1862]])

In [126]:
torch.sigmoid(x)

Out[126]:
tensor([[0.7944, 0.4015], [0.3972, 0.4536]])

In [127]:
计算张量的二次方根

In [128]:
x

Out[128]:
tensor([[1.3517, -0.3991], [-0.4170, -0.1862]])

In [129]:
torch.sqrt(x)

Out[129]:
tensor([[1.1626, nan], [nan, nan]])

In [130]:
# 创建张量
x = torch.arange(10, 10000, 150)
x

Out[130]:
tensor([10, 160, 310, 460, 610, 760, 910, 1060, 1210, 1360, 1510, 1660,
1810, 1960, 2110, 2260, 2410, 2560, 2710, 2860, 3010, 3160, 3310, 3460,
3610, 3760, 3910, 4060, 4210, 4360, 4510, 4660, 4810, 4960, 5110, 5260,
5410, 5560, 5710, 5860, 6010, 6160, 6310, 6460, 6610, 6760, 6910, 7060,
7210, 7360, 7510, 7660, 7810, 7960, 8110, 8260, 8410, 8560, 8710, 8860,
9010, 9160, 9310, 9460, 9610, 9760, 9910])

In[131]:

print(f"Minimum: {x.min()}")
print(f"Maximum: {x.max()}")
# print(f"Mean: {x.mean()}") # 直接求平均会出错
print(f"Mean: {x.type(torch.float32).mean()}") # 需要转换为float32数据类型
print(f"Sum: {x.sum()}")
Minimum: 10
Maximum: 9910
Mean: 4960.0
Sum: 332320

In[132]:

torch.argmax(x),torch.argmin(x)

Out[132]:
(tensor(66), tensor(0))

In[133]:

torch.max(x),torch.min(x)

Out[133]:
(tensor(9910), tensor(10))

In[134]:
# 更改数据类型
y = torch.tensor([[39,339.63],
[36,667.20],
[33,978.07],
[31,897.13],
[29,178.19],
[26,442.25],
[24,314.22],

[21,547.88],
[18,764.25],
[16,588.23],
[13,773.61]],dtype=torch.float32)

In [135]:

y.dtype

Out[135]:

torch.float32

In [136]:

创建一个 float16 类型的张量

tensor_float16 = y.type(torch.float16)
tensor_float16

Out[136]:

tensor([[39.0000, 339.7500], [36.0000, 667.0000], [33.0000, 978.0000], [31.0000, 897.0000], [29.0000, 178.2500], [26.0000, 442.2500], [24.0000, 314.2500], [21.0000, 548.0000], [18.0000, 764.0000], [16.0000, 588.0000], [13.0000, 773.5000]], dtype=torch.float16)

In [137]:

y

Out[137]:

tensor([[39.0000, 339.6300], [36.0000, 667.2000], [33.0000, 978.0700], [31.0000, 897.1300], [29.0000, 178.1900], [26.0000, 442.2500], [24.0000, 314.2200], [21.0000, 547.8800], [18.0000, 764.2500], [16.0000, 588.2300], [13.0000, 773.6100]])

In [138]:

创建一个 int8 类型的张量

tensor_int8 = y.type(torch.int8)
tensor_int8

Out[138]:

tensor([[39, 83], [36, -101], [33, -46], [31, -127], [29, -78], [26, -70], [24, 58], [21, 35], [18, -4], [16, 76], [13, 5]], dtype=torch.int8)

In[139]:

改变视图（与原始数据相同，但更改了视图）

y.view(2,11)

Out[139]:

tensor([[39.0000, 339.6300, 36.0000, 667.2000, 33.0000, 978.0700, 31.0000, 897.1300, 29.0000, 178.1900, 26.0000], [442.2500, 24.0000, 314.2200, 21.0000, 547.8800, 18.0000, 764.2500, 16.0000, 588.2300, 13.0000, 773.6100]])

In[140]:

张量的堆叠

A = torch.arange(10,50,5)
B = torch.arange(20,60,5)

In[141]:

torch.stack([A,B],dim=0)

Out[141]:

tensor([[10, 15, 20, 25, 30, 35, 40, 45], [20, 25, 30, 35, 40, 45, 50, 55]])

In[142]:

torch.stack([A,B],dim=1)

Out[142]:

tensor([[10, 20], [15, 25], [20, 30], [25, 35], [30, 40], [35, 45], [40, 50], [45, 55]])

In[143]:

torch.stack([A,B])

Out[143]:

tensor([[10, 15, 20, 25, 30, 35, 40, 45], [20, 25, 30, 35, 40, 45, 50, 55]])

In[144]:

张量的索引

y = torch.stack([A,B,A,B,A,B,A,B])
y

Out[144]:
tensor([[10, 15, 20, 25, 30, 35, 40, 45], [20, 25, 30, 35, 40, 45, 50, 55], [10, 15, 20, 25, 30, 35, 40, 45], [20, 25, 30, 35, 40, 45, 50, 55], [10, 15, 20, 25, 30, 35, 40, 45], [20, 25, 30, 35, 40, 45, 50, 55], [10, 15, 20, 25, 30, 35, 40, 45], [20, 25, 30, 35, 40, 45, 50, 55]])

In[145]:

获取第0维所有值以及第1维索引为1的值

y[:, 1]

Out[145]:

tensor([15, 25, 15, 25, 15, 25, 15, 25])

In[146]:

D = torch.tensor([[[12,13,14],
                   [15,16,17],
                   [18,19,20]]])

In[147]:

获取第0维和第1维的所有值，但只获取第2维索引为1的值

D[:, :, 1]

Out[147]:

tensor([[13, 16, 19]])

In[148]:

获取第0维的所有值，但仅获取第1维和第2维索引为1的值

D[:, 1, 1]

Out[148]:

tensor([16])

In[149]:

获取第0维和第1维索引为0的值，以及第2维的所有值

D[0, 0, :]   # 与D[0][0]相同

Out[149]:

tensor([12, 13, 14])

In[150]:

D[0][0]

Out[150]:
tensor([12, 13, 14])

根据 CUDA 语义，PyTorch 可以配置为 GPU 运行，具体内容可以参考链接 https://pytorch.org/docs/stable/notes/cuda.html。

In[151]:
```python
# 检查GPU是否可用
import torch
torch.cuda.is_available()
```
Out[151]:

False

In[152]:
```python
# 设置设备类型
device = "cuda" if torch.cuda.is_available() else "cpu"
device
```
Out[152]:

cpu

In[153]:
```python
# 计算设备数量
torch.cuda.device_count()
```
Out[153]:

0

In[154]:
```python
# x = torch.randn(2, 2, device='cpu')    # 在cpu上进行运算
# x = torch.randn(2, 2, device='gpu')    # 在gpu上进行运算
# x = torch.randn(2, 2, device=device)   # 自动选择设备
```

生成随机数的语法不依赖于设备，因此在 CPU 和 GPU 环境中均可正常运行。

In[155]:
```python
# 像numpy中一样对tensor进行扁平化
D.flatten()
```
Out[155]:
tensor([12, 13, 14, 15, 16, 17, 18, 19, 20])

In [156]:
# 纵向连接
cat_rows = torch.cat((A, B), dim=0)
cat_rows

Out[156]:
tensor([10, 15, 20, 25, 30, 35, 40, 45, 20, 25, 30, 35, 40, 45, 50, 55])

In [157]:
A.reshape(2,4)

Out[157]:
tensor([[10, 15, 20, 25], [30, 35, 40, 45]])

In [158]:
B.reshape(2,4)

Out[158]:
tensor([[20, 25, 30, 35], [40, 45, 50, 55]])

In [159]:
cat_cols = torch.cat((A.reshape(2,4), B.reshape(2,4)), dim=1)
cat_cols

Out[159]:
tensor([[10, 15, 20, 25, 20, 25, 30, 35], [30, 35, 40, 45, 40, 45, 50, 55]])

小结

对于已具备 PyTorch 和 Python 编程知识的读者，本章将为您提供一个复习的机会。对于 PyTorch 的新手来说，这些内容则是构建您知识体系的基础要素。在深入探讨更高级的主题之前，确保您对 PyTorch 的术语和基本语法有清晰的理解至关重要。在下一章中，我们将引领您了解如何利用 PyTorch 构建概率模型，包括随机变量的生成、统计分布的应用，以及统计推断的实现。

第 2 章

使用 PyTorch 中的概率分布

在计算图平台如 PyTorch 中，概率和随机变量作为计算模块的核心部分，对于理解其工作原理至关重要。在进行本章的学习之前，需要对概率及相关概念具备一定的认识。在本章中，将深入探讨概率分布的特性，以及如何使用 PyTorch 实现这些分布并对其结果进行解读。

在概率论与统计学领域，随机变量被视为数值结果完全由随机现象决定的变量。其种类繁多，包括正态分布、二项式分布、多项式分布以及伯努利分布等。这些不同的分布类型各具特色，具备各自的特性和适用范围。

PyTorch 的 torch.distributions 模块涵盖了概率分布与采样函数，这些分布类型在计算图中占据着不可或缺的地位。该模块包含了二项式分布、伯努利分布、贝塔分布、分类分布、指数分布、正态分布以及泊松分布等丰富多样的分布类型。这些分布的应用领域广泛，从机器学习模型的构建到深度学习算法的实现，都离不开它们的支持。

秘籍 2-1　采样张量

问题

权重初始化在训练神经网络和各类深度学习模型中占据着至关重要的地位，其中涵盖了卷积神经网络（Convolutional Neural Network，CNN）、深度神经网络（Deep Neural Network，DNN）以及循环神经网络（Recurrent Neural Network，RNN）等模型。而核心问题始终在于如何进行权重的初始化。

解决方案

在权重初始化过程中，有多种方法可供选择，其中包括随机权重初始化。此外，权重初始化可以基于特定的分布进行，包括均匀分布、伯努利分布、多项式分布和正态分布。为了进一步探讨这一话题，本节将重点介绍如何在 PyTorch 中进行权重初始化。通过采取合适的权重初始化策略，有助于提高模型的训练效果和性能。

编程实战

为了运行神经网络，需要将一组初始权重传递至反向传播层，以便计算损失函数并评估模型的准确性。选择合适的初始权重方法至关重要，需根据数据特性、具体任务要求以及所需的优化算法进行权衡。本节将全面探讨各类初始权重的设定方法。

另外，如果需要确保实验结果的可重复性，应手动设定随机种子值，以确保每次实验的一致性。

```
In[1]:
import torch
In[2]:
print(torch.cuda.is_available())
False
```

In[3]:
CUDA 是由 NVIDIA 开发的 API，旨在提供 GPU 的接口

In[4]:
```
x = torch.randn(10)
print(x.device)
cpu
```

In[5]:
对张量进行随机抽样

In[6]:
```
torch.manual_seed(1234)
```

In[7]:
```
torch.manual_seed(1234)
torch.randn(4,4)
```

Out[7]:
tensor([[-0.1117, -0.4966, 0.1631, -0.8817], [0.0539, 0.6684, -0.0597, -0.4675], [-0.2153, 0.8840, -0.7584, -0.3689], [-0.3424, -1.4020, 0.3206, -1.0219]])

随机种子值可由用户自定义，确保随机数完全基于偶然性产生。此外，随机数也可通过统计分布生成。连续均匀分布的概率密度函数遵循以下公式定义：

$$f(x) = \begin{cases} \dfrac{1}{b-a}, & a \leq x \leq b \\ 0, & x < a \text{ 或 } x > b \end{cases}$$

在关于 x 的函数上有两个点 a 和 b，其中 a 是起点，b 是终点。在连续均匀分布中，每个数字被选中的机会是均等的。在以下示例中，设定起点为 0，终点为 1；在这两个数字之间，随机选择 16 个元素。

In[8]:
从统计分布中生成随机数

In[9]:
```
torch.Tensor(4, 4).uniform_(0, 1)    # 均匀分布中的随机数
```

Out[9]:
tensor([[0.2837, 0.6567, 0.2388, 0.7313], [0.6012, 0.3043, 0.2548, 0.6294], [0.9665, 0.7399, 0.4517, 0.4757], [0.7842, 0.1525, 0.6662, 0.3343]])

在统计学中,伯努利分布被视为一种离散概率分布。该分布具有两种可能的结果,即事件发生时取值为 1,事件未发生时取值为 0。

在离散概率分布的情况下,需要计算的是概率质量函数,而不是概率密度函数。概率质量函数的定义如下:

$$\begin{cases} q = (1-p), & k = 0 \\ p, & k = 1 \end{cases}$$

使用伯努利分布,可以构造一个 4×4 的均匀分布矩阵格式,以生成样本张量,如下所示:

In[10]:
现在假设输入的张量是概率值,应用伯努利分布

In[11]:
torch.bernoulli(torch.Tensor(4, 4).uniform_(0, 1))

Out[11]:
tensor([[0., 0., 0., 0.], [1., 0., 1., 0.], [1., 0., 1., 1.], [0., 0., 0., 0.]])

以下是如何在多项式分布中生成样本随机值的方法:在多项式分布中,采样方式可以选择有放回或无放回。默认情况下,multinomial 函数采用无放回的方式进行采样,并将结果作为张量的索引位置返回。若需采用有放回的方式进行采样,则需要在采样时进行相应设定。

In[12]:
从多项式分布中进行采样

In[13]:
torch.Tensor([10, 10, 13, 10,34,45,65,67,87,89,87,34])

Out[13]:
tensor([10., 10., 13., 10., 34., 45., 65., 67., 87., 89., 87., 34.])

In[14]:
torch.multinomial(torch.tensor([10., 10., 13., 10.,
                                34., 45., 65., 67.,
                                87., 89., 87., 34.]),
                  3)

Out[14]:
tensor([4, 5, 7])

在多项式分布中进行有放回的采样，将返回对应张量的索引值。

In[15]:

```
torch.multinomial(torch.tensor([10., 10., 13., 10.,
                                34., 45., 65., 67.,
                                87., 89., 87., 34.]),
                  5, replacement=True)
```

Out[15]:

tensor([10, 5, 9, 10, 5])

在训练神经网络、深度神经网络、卷积神经网络和循环神经网络时，利用正态分布进行权重初始化是一种常见的方法。以下是利用正态分布生成一组随机权重的过程。

In[16]:

从正态分布中生成随机数

In[17]:

```
torch.normal(mean=torch.arange(1., 11.),
             std=torch.arange(1, 0, -0.1))
```

Out[17]:

tensor([1.5236, 2.2441, 2.7375, 3.9521, 5.4380, 5.5158, 8.2489, 8.1645, 9.0575, 9.8627])

In[18]:

```
torch.normal(mean=0.5,
             std=torch.arange(1., 6.))
```

Out[18]:

tensor([1.1144, 0.0361, 1.2766, -1.3999, -0.1648])

In[19]:

```
torch.normal(mean=0.5,
             std=torch.arange(0.2,0.6))
```

Out[19]:

tensor([-0.0844])

In[20]:

计算描述性统计数据：平均数

```
torch.mean(torch.tensor([10., 10., 13., 10., 34.,
                         45., 65., 67., 87., 89., 87., 34.]))
```

Out[20]:

tensor(45.9167)

In[21]:

行均值和列均值

d = torch.randn(4, 5)

d

Out[21]:

tensor([[-1.6406, 0.9295, 1.2907, 0.2612, 0.9711], [0.3551, 0.8562, -0.3635, -0.1552, -1.2282], [1.2445, 1.1750, -0.2217, -2.0901, -1.2658], [-1.8761, -0.6066, 0.7470, 0.4811, 0.6234]])

In[22]:

torch.mean(d,dim=0)

Out[22]:

tensor([-0.4793, 0.5885, 0.3631, -0.3757, -0.2249])

In[23]:

torch.mean(d,dim=1)

Out[23]:

tensor([0.3624, -0.1071, -0.2316, -0.1262])

In[24]:

计算中位数

torch.median(d,dim=0)

Out[24]:

torch.return_types.median(values=tensor([-1.6406, 0.8562, -0.2217, -0.1552, -1.2282]), indices=tensor([0, 1, 2, 1, 1]))

In[25]:

torch.median(d,dim=1)

Out[25]:

torch.return_types.median(values=tensor([0.9295, -0.1552, -0.2217, 0.4811]), indices=tensor([1, 3, 2, 3]))

秘籍 2-2 可变张量

问题

什么是 PyTorch 中的变量？如何定义它？什么是 PyTorch 中的随机变量？

解决方案

在 PyTorch 中，算法是通过计算图进行表示的。变量这一概念与张量对象、对应的梯度和构建函数引用密切相关。为了便于理解，可以将梯度视为函数的斜率。具体来说，函数的斜率可以通过对参数进行求导来获得。以线性回归（Y = W*X + alpha）为例，该算法的变量计算图如图 2-1 所示。

在 PyTorch 中，变量是计算图中的节点，用于存储数据和梯度。在训练神经网络模型时，每次迭代后，需要计算损失函数相对于模型参数（如权重和偏置）的梯度。然后，通常使用梯度下降算法来更新权重。图 2-1 演示了如何在 PyTorch 框架中使用神经网络模型来实现线性回归方程。

在处理图结构计算时，任务顺序和排序的准确性至关重要。图 2-1 中涉及的 X、Y、W 和 alpha 均为一维张量。在实施反向传播调整权重以匹配 Y 时，务必注意箭头的指向变化，以确保 Y 与预测的 Y′ 之间的误差或损失函数达到最小值。

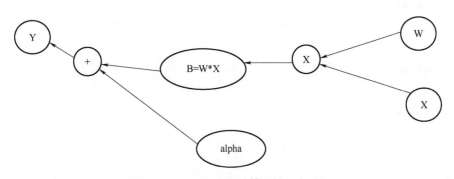

图 2-1　PyTorch 实现计算图的一个示例

编程实战

以下是一个示例脚本,展示了如何利用变量构建计算图。在此脚本中,创建了三个张量类型的变量对象:x1、x2 和 x3。这些变量是从 $a=12$ 到 $b=23$ 之间随机生成的数值点。该计算图仅涉及乘法和加法两种运算,并可以计算出具有梯度的最终结果。

在 PyTorch 中,我们通过 Autograd 模块来计算神经网络模型中权重和偏置的偏导数。这个模块的变量是专门设计的,能够在神经网络模型运行反向传播过程中,保存模型参数的变化值。这些变量本质上只是张量的包装器,它们具有三个关键属性:数据、梯度和函数。

```
In [26]:
from torch.autograd import Variable
In [27]:
Variable(torch.ones(2,2),requires_grad=True)
Out[27]:
tensor([[1., 1.], [1., 1.]], requires_grad=True)
In [28]:
a, b = 12,23
x1 = Variable(torch.randn(a,b),
              requires_grad=True)
x2 = Variable(torch.randn(a,b),
              requires_grad=True)
x3 = Variable(torch.randn(a,b),
              requires_grad=True)
In [29]:
c = x1 * x2
d = a + x3
e = torch.sum(d)

e.backward()

print(e)
tensor(3278.1235, grad_fn=<SumBackward0>)
```

In[30]:

x1.data

Out[30]:

tensor([[-4.9545e-02, 6.2245e-01, 1.6573e-01, 3.1583e-01, 2.4915e-01,
-4.9784e-01, 2.9079e+00, 1.6201e+00, -6.4459e-01, -1.9885e-02, 1.6222e+00,
1.4239e+00, 9.0691e-01, 7.6310e-02, 1.1225e+00, -1.2433e+00, -6.7258e-01,
8.8433e-01, -6.6589e-01, -7.3347e-01, -2.7599e-01, 5.5485e-01,
-1.9303e+00],................

In[31]:

x2.data

Out[31]:

tensor([[-7.5597e-01, -1.1689e+00, -9.3890e-01, 8.8566e-01, 1.3764e+00,
-7.8276e-01, 2.2200e-01, 7.3758e-02, -6.9147e-01, -5.1308e-01, 1.1427e+00,
-1.0126e+00, 1.1602e-01, -1.0350e+00, 1.0803e+00, -7.9977e-01,
-9.1219e-02, 5.0242e-01, -4.5173e-01, -4.8067e-01, 5.9066e-01, 1.6343e-01,
-3.1368e-02],................

In[32]:

x3.data

Out[32]:

tensor([[0.2499, 0.2458, 0.1029, -0.6494, -0.3258, 0.8149, 0.4049, 0.2481,
0.4841, 0.3293, -1.2471, 0.2117, 1.4315, 0.0502, -0.3668, 0.8378, -0.7901,
0.0267, -0.3120, 2.4534, 0.7926, 0.2382, -0.5245]……..

秘籍 2-3　统计学基础

问题

如何从 Torch 张量中计算基本统计信息（如平均值、中位数、众数等）？

解决方案

在处理数据时，使用 PyTorch 进行基本统计量的计算可以结合概率分布和统计测试来进行推理。尽管 Torch 的功能与 Numpy 相似，但 Torch 函数具备 GPU 加速的优势。本节将演示如何利用 PyTorch 创建基本统计量。

编程实战

在处理一维张量时，平均值计算是一个相对简单的任务；然而，对于二维张量，情况则更为复杂。在进行平均值、中位数或众数的计算时，需要额外传递一个参数，用以指明在哪个维度上进行计算。

In[33]:
```
# 计算描述性统计量：平均值
torch.mean(torch.tensor([10., 10., 13., 10., 34.,
                         45., 65., 67., 87., 89., 87., 34.]))
```

Out[33]:

tensor(45.9167)

In[34]:
```
# 行间和列间的平均值
d = torch.randn(4, 5)
d
```

Out[34]:

tensor([[-1.6406, 0.9295, 1.2907, 0.2612, 0.9711], [0.3551, 0.8562, -0.3635, -0.1552, -1.2282], [1.2445, 1.1750, -0.2217, -2.0901, -1.2658], [-1.8761, -0.6066, 0.7470, 0.4811, 0.6234]])

In[35]:

torch.mean(d,dim=0)

Out[35]:

tensor([-0.4793, 0.5885, 0.3631, -0.3757, -0.2249])

In[36]:

torch.mean(d,dim=1)

Out[36]:

tensor([0.3624, -0.1071, -0.2316, -0.1262])

中位数、众数以及标准差的计算均可采用相同的方法进行编写。

In[37]:
# 计算中位数
torch.median(d,dim=0)

Out[37]:
torch.return_types.median(values=tensor([-1.6406, 0.8562, -0.2217, -0.1552, -1.2282]), indices=tensor([0, 1, 2, 1, 1]))

In[38]:
torch.median(d,dim=1)

Out[38]:
torch.return_types.median(values=tensor([0.9295, -0.1552, -0.2217, 0.4811]), indices=tensor([1, 3, 2, 3]))

In[39]:
# 计算众数
torch.mode(d)

Out[39]:
torch.return_types.mode(values=tensor([-1.6406, -1.2282, -2.0901, -1.8761]), indices=tensor([0, 4, 3, 0]))

In[40]:
torch.mode(d,dim=0)

Out[40]:
torch.return_types.mode(values=tensor([-1.8761, -0.6066, -0.3635, -2.0901, -1.2658]), indices=tensor([3, 3, 1, 2, 2]))

In[41]:
torch.mode(d,dim=1)

Out[41]:
torch.return_types.mode(values=tensor([-1.6406, -1.2282, -2.0901, -1.8761]), indices=tensor([0, 4, 3, 0]))

  标准差用于衡量数据与中心趋势的偏离程度，能够反映数据的一致性。通过标准差，可以判断数据是否存在显著的波动。

In[42]:
# 计算标准差
torch.std(d)

Out[42]:

tensor(1.0944)

In[43]:

torch.std(d,dim=0)

Out[43]:

tensor([1.5240, 0.8083, 0.7911, 1.1730, 1.1889])

In[44]:

torch.std(d,dim=1)

Out[44]:

tensor([1.1807, 0.7852, 1.4732, 1.1165])

In[45]:

计算方差

torch.var(d)

Out[45]:

tensor(1.1978)

In[46]:

torch.var(d,dim=0)

Out[46]:

tensor([2.3224, 0.6534, 0.6259, 1.3758, 1.4134])

In[47]:

torch.var(d,dim=1)

Out[47]:

tensor([1.3940, 0.6166, 2.1703, 1.2466])

In[48]:

计算最小值和最大值

torch.min(d)

Out[48]:

tensor(-2.0901)

In[49]:

torch.min(d,dim=0)

Out[49]:

torch.return_types.min(values=tensor([-1.8761, -0.6066, -0.3635, -2.0901, -1.2658]), indices=tensor([3, 3, 1, 2, 2]))

In[50]:

torch.min(d,dim=1)

Out[50]:

torch.return_types.min(values=tensor([-1.6406, -1.2282, -2.0901, -1.8761]), indices=tensor([0, 4, 3, 0]))

In[51]:

torch.max(d)

Out[51]:

tensor(1.2907)

In[52]:

torch.max(d,dim=0)

Out[52]:

torch.return_types.max(values=tensor([1.2445, 1.1750, 1.2907, 0.4811, 0.9711]), indices=tensor([2, 2, 0, 3, 0]))

In[53]:

torch.max(d,dim=1)

Out[53]:

torch.return_types.max(values=tensor([1.2907, 0.8562, 1.2445, 0.7470]), indices=tensor([2, 1, 0, 2])

In[54]:

对张量进行排序

torch.sort(d)

Out[54]:

torch.return_types.sort(values=tensor([[-1.6406, 0.2612, 0.9295, 0.9711, 1.2907], [-1.2282, -0.3635, -0.1552, 0.3551, 0.8562], [-2.0901, -1.2658, -0.2217, 1.1750, 1.2445], [-1.8761, -0.6066, 0.4811, 0.6234, 0.7470]]), indices=tensor([[0, 3, 1, 4, 2], [4, 2, 3, 0, 1], [3, 4, 2, 1, 0], [0, 1, 3, 4, 2]]))

In[55]:

torch.sort(d,dim=0)

Out[55]:

torch.return_types.sort(values=tensor([[-1.8761, -0.6066, -0.3635, -2.0901, -1.2658], [-1.6406, 0.8562, -0.2217, -0.1552, -1.2282], [0.3551, 0.9295, 0.7470, 0.2612, 0.6234], [1.2445, 1.1750, 1.2907, 0.4811, 0.9711]]), indices=tensor([[3, 3, 1, 2, 2], [0, 1, 2, 1, 1], [1, 0, 3, 0, 3], [2, 2, 0, 3, 0]]))

In[56]:

torch.sort(d,dim=0,descending=True)

Out[56]:

torch.return_types.sort(values=tensor([[1.2445, 1.1750, 1.2907, 0.4811, 0.9711], [0.3551, 0.9295, 0.7470, 0.2612, 0.6234], [-1.6406, 0.8562, -0.2217, -0.1552, -1.2282], [-1.8761, -0.6066, -0.3635, -2.0901, -1.2658]]), indices=tensor([[2, 2, 0, 3, 0], [1, 0, 3, 0, 3], [0, 1, 2, 1, 1], [3, 3, 1, 2, 2]]))

In[57]:

torch.sort(d,dim=1,descending=True)

Out[57]:

torch.return_types.sort(values=tensor([[1.2907, 0.9711, 0.9295, 0.2612, -1.6406], [0.8562, 0.3551, -0.1552, -0.3635, -1.2282], [1.2445, 1.1750, -0.2217, -1.2658, -2.0901], [0.7470, 0.6234, 0.4811, -0.6066, -1.8761]]), indices=tensor([[2, 4, 1, 3, 0], [1, 0, 3, 2, 4], [0, 1, 2, 4, 3], [2, 4, 3, 1, 0]]))

In[58]:

from torch.autograd import Variable

In[59]:

Variable(torch.ones(2,2),requires_grad=True)

Out[59]:

tensor([[1., 1.], [1., 1.]], requires_grad=True)

In[60]:

```
a, b = 12,23
x1 = Variable(torch.randn(a,b),
              requires_grad=True)
x2 = Variable(torch.randn(a,b),
              requires_grad=True)
```

```
x3 =Variable(torch.randn(a,b),
            requires_grad=True)
```

In [61]:
```
c = x1 * x2
d = a + x3
e = torch.sum(d)

e.backward()

print(e)

tensor(3278.1235, grad_fn=<SumBackward0>)
```

秘籍 2-4　梯度计算

问题

如何使用 PyTorch 从样本张量中计算基本梯度？

解决方案

在处理样本数据集 datase0074 时，关注两个变量 x 和 y。基于给定的初始权重，能否计算每次迭代后的梯度呢？接下来，将通过一个实例来详细探讨这一问题。

编程实战

x_data 和 y_data 是列表数据类型。要计算这两个数据集的梯度，需要经过几个关键步骤：首先，计算损失函数；其次，进行前向传播过程；最后，执行循环训练。

在前向传播阶段，通过计算权重张量与输入张量的矩阵乘积来得出前向函数的输出结果。

In [62]:
```
from torch import FloatTensor
from torch.autograd import Variable

a = Variable(FloatTensor([5]))
weights = [Variable(FloatTensor([i]), requires_grad=True) for i in (12, 53, 91, 73)]
w1, w2, w3, w4 = weights
b = w1 * a
c = w2 * a
d = w3 * b + w4 * c
Loss = (10 - d)
Loss.backward()

for index, weight in enumerate(weights, start=1):
    gradient, *_ = weight.grad.data
    print(f"Gradient of w{index} w.r.t to Loss: {gradient}")
Gradient of w1 w.r.t to Loss: -455.0
Gradient of w2 w.r.t to Loss: -365.0
Gradient of w3 w.r.t to Loss: -60.0
Gradient of w4 w.r.t to Loss: -265.0
```

In [63]:
```
# 使用前向传播
def forward(x):
    return x * w
```

In [64]:
```
import torch
from torch.autograd import Variable

x_data = [11.0, 22.0, 33.0]
y_data = [21.0, 14.0, 64.0]

w = Variable(torch.Tensor([1.0]), requires_grad=True) # 任何随机值
# 在训练之前
print("predict (before training)", 4, forward(4).data[0])
predict (before training) 4 tensor (4.)
```

In [65]:
```
# 定义损失函数
```

```python
def loss(x, y):
    y_pred = forward(x)
    return (y_pred - y) * (y_pred - y)
```

In [66]:

```python
# 运行训练循环
for epoch in range(10):
    for x_val, y_val in zip(x_data, y_data):
        l = loss(x_val, y_val)
        l.backward()
        print("\tgrad: ", x_val, y_val, w.grad.data[0])
        w.data = w.data - 0.01 * w.grad.data

        # 权重更新后手动将梯度设为零
        w.grad.data.zero_()

    print("progress:", epoch, l.data[0])
```

grad: 11.0 21.0 tensor(-220.)
  grad: 22.0 14.0 tensor(2481.6001)
  grad: 33.0 64.0 tensor(-51303.6484)
progress: 0 tensor(604238.8125)……………………

In [67]:

```python
# 训练后
print("predict (after training)", 4, forward(4).data[0])
```
predict (after training) 4 tensor(-9.2687e+24)

  在以下程序片段中，展示了如何利用张量中的 Variable 方法从损失函数中计算梯度。

```python
a = Variable(FloatTensor([5]))
weights = [Variable(FloatTensor([i]), requires_grad=True) for i in (12, 53, 91, 73)]
w1, w2, w3, w4 = weights
b = w1 * a
c = w2 * a
d = w3 * b + w4 * c
Loss = (10 - d)
Loss.backward()
```

秘籍 2-5　张量运算之一

问题

如何基于 Variable 执行如矩阵乘法这样的计算或运算？

解决方案

在 Pytorch 中，张量被封装在 Variable 类中，并具有三个属性：grad、volatile 和 gradient。

编程实战

在实现权重更新过程时，创建 Variable 并提取其属性是至关重要的，因为这涉及梯度的计算。为了执行矩阵乘法运算，可以利用 mm 模块。

In[68]:
```
x = Variable(torch.Tensor(4, 4).uniform_(-4, 5))
y = Variable(torch.Tensor(4, 4).uniform_(-3, 2))
# 矩阵乘法
z = torch.mm(x, y)
print(z.size())
torch.Size([4, 4])
```

以下程序展示了 Variable 的属性，Variable 是张量包装器。

In[69]:
```
z = Variable(torch.Tensor(4, 4).uniform_(-5, 5))
print(z)
tensor([[-0.3071, -3.6691, -2.8417, -1.1818],
        [-1.4654, -0.4344, -2.0130, -2.3842],
        [ 1.3962,  1.4962, -2.0996,  1.8881],
        [-1.9797,  0.2337, -1.0308,  0.1266]])
```

In [70]:
```
print('Requires Gradient : %s ' % (z.requires_grad))
print('Volatile : %s ' % (z.volatile))
print('Gradient : %s ' % (z.grad))
print(z.data)
Requires Gradient : False
Volatile : False
Gradient : None
tensor([[-0.3071, -3.6691, -2.8417, -1.1818],
        [-1.4654, -0.4344, -2.0130, -2.3842],
        [ 1.3962,  1.4962, -2.0996,  1.8881],
        [-1.9797,  0.2337, -1.0308,  0.1266]])
```

秘籍 2-6　张量运算之二

问题

如何基于 Variable 执行如矩阵与向量的运算、矩阵与矩阵的运算以及向量与向量的运算？

解决方案

矩阵运算成功的必要条件之一是确保张量的长度相匹配或兼容，以便正确执行代数表达式。

编程实战

在数学中，标量的张量定义仅指一个具体的数值。对于一维张量，它等同于向量；对于二维张量，它相当于矩阵。当张量的维度扩展到更高层次时，将其统称为张量。在进行代数运算时，在 PyTorch 中参与运算的矩阵、向量或标量的维度应当保持一致，以确保计算的准确性和合法性。

In[71]:

张量操作

In[72]:

mat1 = torch.FloatTensor(4,4).uniform_(0,1)
mat1

Out[72]:

tensor([[0.9002, 0.9188, 0.1386, 0.3701], [0.1947, 0.2268, 0.9587, 0.2615], [0.7256, 0.7673, 0.5667, 0.1863], [0.4642, 0.4016, 0.9981, 0.8452]])

In[73]:

mat2 = torch.FloatTensor(4,4).uniform_(0,1)
mat2

Out[73]:

tensor([[0.4962, 0.4947, 0.8344, 0.6721], [0.1182, 0.5997, 0.8990, 0.8252], [0.1466, 0.1093, 0.8135, 0.9047], [0.2486, 0.1873, 0.6159, 0.2471]])

In[74]:

vec1 = torch.FloatTensor(4).uniform_(0,1)
vec1

Out[74]:

tensor([0.7582, 0.6879, 0.8949, 0.3995])

In[75]:

标量加法

In[76]:

mat1 + 10.5

Out[76]:

tensor([[11.4002, 11.4188, 10.6386, 10.8701], [10.6947, 10.7268, 11.4587, 10.7615], [11.2256, 11.2673, 11.0667, 10.6863], [10.9642, 10.9016, 11.4981, 11.3452]])

In[77]:

标量减法

In[78]:

mat2 - 0.20

Out[78]:

tensor([[0.2962, 0.2947, 0.6344, 0.4721], [-0.0818, 0.3997, 0.6990,

0.6252], [-0.0534, -0.0907, 0.6135, 0.7047], [0.0486, -0.0127, 0.4159, 0.0471]])

In[79]:

向量和矩阵相加

In[80]:

mat1 + vec1

Out[80]:

tensor([[1.6584, 1.6067, 1.0335, 0.7695], [0.9530, 0.9147, 1.8537, 0.6610], [1.4839, 1.4553, 1.4616, 0.5858], [1.2224, 1.0895, 1.8931, 1.2446]])

In[81]:

mat2 + vec1

Out[81]:

tensor([[1.2544, 1.1826, 1.7293, 1.0716], [0.8764, 1.2876, 1.7939, 1.2247], [0.9049, 0.7972, 1.7084, 1.3042], [1.0068, 0.8752, 1.5108, 0.6466]])

如果 mat1 和 mat2 的维度不同，则它们不适合矩阵加法或乘法；如果维度相同，则可以将其相乘。在下面的脚本中，如果将不匹配的维度相乘时，矩阵加法会抛出错误。

In[82]:

矩阵加法

In[83]:

mat1 + mat2

Out[83]:

tensor([[1.3963, 1.4135, 0.9730, 1.0422], [0.3129, 0.8265, 1.8577, 1.0867], [0.8722, 0.8766, 1.3802, 1.0910], [0.7127, 0.5888, 1.6141, 1.0923]])

In[84]:

mat1 * mat1

Out[84]:

tensor([[0.8103, 0.8442, 0.0192, 0.1370], [0.0379, 0.0514, 0.9192, 0.0684], [0.5265, 0.5888, 0.3211, 0.0347], [0.2155, 0.1613, 0.9963, 0.7143]])

秘籍 2-7　统计分布

问题

了解统计分布的相关知识，对于神经网络中的权重归一化、权重初始化以及梯度计算等关键环节至关重要。在具体应用时，应针对不同情况选择合适的分布。

解决方案

在各种问题场景中，通常会选择使用那些最为常用的统计分布，并为其设定适当的参数。这些统计分布的选用，均遵循了预先建立的数学公式，确保了统计结果的准确性和可靠性。

编程实战

在概率论中，**伯努利分布**是**二项式分布**的一种特例，其试验次数限定为一次。不同于二项式分布，伯努利分布仅考虑单个试验的结果，即成功或失败。在伯努利分布中，随机变量只取 0 或 1 两个离散值，对应于事件的失败和成功。以抛硬币为例，若将正面视为成功，反面视为失败，那么每一次抛硬币的结果都符合伯努利分布。接下来，将演示如何在程序中实现伯努利分布。

```
In[85]:
# 伯努利分布
In[86]:
from torch.distributions.bernoulli import Bernoulli
In[87]:
dist = Bernoulli(torch.tensor([0.3,0.6,0.9]))
In[88]:
dist.sample()   # 样本是二进制的，当 p 为真时取 1，当 p 为假时取 0
```

Out[88]:

tensor([0., 1., 0.])

In[89]:

以事件概率为参数,创建一个伯努利分布

样本是二进制的(0 或 1)。它们以概率 p 取值为 1,以概率 1-p 取值为 0

贝塔分布是一种在 0 到 1 范围内定义的连续随机变量族,广泛应用于贝叶斯推断分析中。该分布通过使用先验信息和样本数据来更新和修正对未知参数的推断,对于解决多种统计问题具有重要意义。

In[90]:

from torch.distributions.beta import Beta

In[91]:

dist = Beta(torch.tensor([0.5]), torch.tensor([0.5]))
dist

Out[91]:

Beta()

In[92]:

dist.sample()

Out[92]:

tensor([0.8935])

在满足二分结果且实验可重复的条件下,应采用二项式分布进行概率分析。二项式分布属于离散概率分布类别,其中成功的概率为 1,而失败的概率为 0。该分布用于模拟多次试验中成功事件的数量。

In[93]:

from torch.distributions.binomial import Binomial

In[94]:

dist = Binomial(100, torch.tensor([0 , .2, .8, 1]))

In[95]:

dist.sample()

Out[95]:

tensor([0., 21., 83., 100.])

In[96]:

#100 为试验次数

#0、0.2、0.8 和 1 是事件概率

在概率论和统计学领域,分类分布可以被定义为一种广义伯努利分布。这种离散概率分布主要用于解释随机变量可能取某一个可能类别的结果,且每个类别的概率独立地在张量中指定。

In[97]:

```
from torch.distributions.categorical import Categorical
```

In[98]:

```
dist = Categorical(torch.tensor([ 0.20, 0.20, 0.20, 0.20, 0.20 ]))
dist
```

Out[98]:

```
Categorical(probs: torch.Size([5]))
```

In[99]:

```
dist.sample()
```

Out[99]:

```
tensor(2)
```

In[100]:

#0.20, 0.20, 0.20, 0.20, 0.20 为事件概率

拉普拉斯分布又称为双指数分布,是一种连续的概率分布函数。在语音识别系统中,该分布用于理解先验概率。而在贝叶斯回归中,它也可以有效地确定先验概率。

拉普拉斯分布包含 loc 和 scale 两个参数。参数 loc 也称为平均值或位置参数,而参数 scale 则是标准差参数。

In[101]:

#Laplace 分布包含 Loc 和 scale 两个参数

In[102]:

```
from torch.distributions.laplace import Laplace
```

In[103]:

```
dist = Laplace(torch.tensor([10.0]), torch.tensor([0.990]))
dist
```

Out[103]:

Laplace(loc: tensor([10.]), scale: tensor([0.9900]))

In[104]:

dist.sample()

Out[104]:

tensor([9.6554])

正态分布的实用性源自其中心极限定理的特性，使其成为众多统计分析中不可或缺的部分。该分布通过平均值和标准差进行定义，一旦明确了这两个参数，便能对事件的概率进行有效的评估。正态分布如图 2-2 所示。

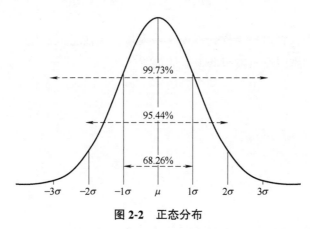

图 2-2　正态分布

In[105]:

正态(高斯)分布包含 Loc 和 scale 两个参数

In[106]:

from torch.distributions.normal import Normal

In[107]:

dist = Normal(torch.tensor([100.0]), torch.tensor([10.0]))
dist

Out[107]:

Normal(loc: tensor([100.]), scale: tensor([10.]))

In[108]:

dist.sample()

Out[108]:

tensor([84.3435])

小结

在本章中，我们深入探讨了采样分布和如何从分布中随机生成数值。神经网络作为张量操作的核心应用，其重要性不言而喻。对于任何机器学习或深度学习模型的实现，梯度计算、权重更新、偏差计算以及持续的偏差更新都是不可或缺的步骤。

此外，本章还对 PyTorch 所支持的统计分布进行了详述，并明确了每种分布的应用场景。

在下一章中，将进一步探究深度学习模型，包括卷积神经网络、循环神经网络、深度神经网络以及自编码器模型的细节。

第 3 章

使用 PyTorch 中的卷积神经网络和循环神经网络

在本章中,将深入探讨卷积神经网络和循环神经网络的原理及实现,并重点解析在 PyTorch 框架中的具体操作。卷积神经网络在图像分类、目标检测和物体识别等领域有广泛的应用,而在大规模图像分类模型的构建与优化方面,PyTorch 框架则展现出显著的速度优势。另外,循环神经网络在自然语言处理任务中发挥着关键作用,涵盖了文本分类、情感分析、主题提取以及音频识别等多个子领域。本章将详述如何构建卷积神经网络和循环神经网络模型,选择合适的优化器,保存和加载已训练模型,以及如何利用模型进行预测。

秘籍 3-1 设置损失函数

问题

如何设置损失函数并进行优化?选择合适的损失函数能够提高模型收敛的可靠性。

解决方案

在本节中,将定义一个新的张量作为更新变量,同时将张量数据输入样例模型,并计算误差或损失。接着,通过计算损失函数中的变化率,以评估所选损失函数在模型收敛过程中的效果。

编程实战

在给定的示例中,t_c 和 t_u 是两个张量,可以用任意两个 NumPy 数组来构建。

In[1]:
import torch

In[2]:
torch.__version__

Out[2]:
'1.12.1+cu113'

In[3]:
torch.tensor

Out[3]:
<function torch._VariableFunctionsClass.tensor>

In[4]:
t_c = torch.tensor([0.5, 14.0, 15.0, 28.0, 11.0, 8.0, 3.0, -4.0, 6.0, 13.0, 21.0])
t_u = torch.tensor([35.7, 55.9, 58.2, 81.9, 56.3, 48.9, 33.9, 21.8, 48.4, 60.4, 68.4])

在演示计算过程中,采用样例模型来展示线性方程的解法。该模型所采用的损失函数为均方误差(Mean Squared Error,MSE)。随着章节的深入,将逐步提升模型的复杂性。当前阶段,仅涉及一个相对简单的线性方程计算。

In[5]:
# 人的身高
t_c = torch.tensor([58.0, 59.0, 60.0, 61.0, 62.0, 63.0, 64.0, 65.0, 66.0, 67.0, 68.0, 69.0, 70.0, 71.0, 72.0])

In[6]:
# 人的体重
t_u = torch.tensor([115.0, 117.0, 120.0, 123.0, 126.0, 129.0, 132.0, 135.0,
139.0, 142.0, 146.0, 150.0, 154.0, 159.0,164.0])

在接下来的步骤中，将对模型进行定义。这里涉及一个重要的参数 w，它被定义为权重张量。这个权重张量将与另一个张量 t_u 进行相乘操作。经过这一乘法运算后，结果会再与一个常量张量 b 进行加法运算。对于损失函数我们选择了一个自定义函数。当然，也可以直接调用 PyTorch 内置的预定义损失函数。在下面的示例中，t_u 表示输入数据所对应的张量，t_p 则是模型的预测输出（即预测的张量），而 t_c 则是预先计算得出的真实值张量。在计算损失函数时，将 t_p 与 t_c 进行对比。

In[7]:
```
def model(t_u, w, b):
    return w * t_u + b
```

In[8]:
```
def loss_fn(t_p, t_c):
    squared_diffs = (t_p - t_c)**2
    return squared_diffs.mean()
```

在张量运算中，公式 w*t_u + b 被用来表示线性方程。这个方程基于张量进行计算，反映数据的内在关系。

In[9]:
```
w = torch.ones(1)
b = torch.zeros(1)

t_p = model(t_u, w, b)
t_p
```
Out[9]:
tensor([115., 117., 120., 123., 126., 129., 132., 135., 139., 142., 146.,
150., 154., 159., 164.])

In[10]:
```
loss = loss_fn(t_p, t_c)
loss
```
Out[10]:
tensor(5259.7334)

初始损失值为 5259.7334，这个数值较高是由于第一轮选择的权重过大所致。在迭代过程中，误差会反向传播以降低第二轮的误差，为此需要更新初始权重集。因此，损失函数的变化率在估计过程中的权重更新中起到了至关重要的作用。

In [11]:
```
delta = 0.1

loss_rate_of_change_w = (loss_fn(model(t_u,
                                       w + delta, b),
                                 t_c) - loss_fn(model(t_u, w - delta, b),
                                                t_c)) / (2.0 * delta)
```

In [12]:
```
learning_rate = 1e-2
w = w - learning_rate * loss_rate_of_change_w
```

In [13]:
```
loss_rate_of_change_b = (loss_fn(model(t_u, w, b + delta), t_c) -
                         loss_fn(model(t_u, w, b - delta), t_c)) /
                        (2.0 * delta)
b = b - learning_rate * loss_rate_of_change_b
```

In [14]:
```
b
```
Out [14]:
```
tensor([544.])
```

在优化损失函数的过程中，有两个参数可以更新损失函数的变化率：当前迭代的损失率和前一轮迭代的损失率。为了确保模型收敛，需要密切监测这两者之间的差值。如果差值超过设定的阈值，即 delta 值，就需要更新权重张量。在当前的脚本中，delta 和学习率都是预设的静态值，用户可以根据实际需求进行调整。

以下是一个关于二维张量中均方损失函数应用的简单示例，该张量的大小为 10×5。在此示例中，使用 PyTorch 神经网络模块中的内置函数 MSELoss 来计算均方误差损失。

In [15]:
```
from torch import nn
loss = nn.MSELoss()
```

```
input = torch.randn(10, 5, requires_grad=True)
target = torch.randn(10, 5)
output = loss(input, target)
output.backward()
```

在输出用于反向传播的梯度计算时,显示为 MSELoss。

In [16]:
output.grad_fn

Out[16]:
<MseLossBackward0 at 0x7f424abd0f50>

秘籍 3-2　估计损失函数的导数

问题

如何估计损失函数的导数?

解决方案

在以下示例中,将损失函数更改为输入张量与输出张量之间差异的两倍,而不是使用均方误差损失函数(MSELoss)。此外,还定义了一个名为 grad_fn 的自定义函数,用于计算损失函数的导数。该函数接受输入张量和输出张量作为输入,并返回它们之间的梯度。

编程实战

我们注意到,在上一节中,脚本输出的最后一行显示了 grad_fn,它被嵌入在输出对象张量中。现在,将详细解释损失函数导数的计算过程。这里需要明确的是,grad_fn 实际上代表了损失函数相对于模型参数的导数。这正是将在下面的 grad_fn 函数中所实现的功能。

In [17]:
input

Out[17]:

tensor([[-1.0665, -0.8880, -1.1156, -0.1595, 0.2342], [3.1369, -0.6062, 1.0556, 1.9240, 1.0309], [1.8270, -0.8902, 1.8918, 1.0523, 1.8231], [0.0969, 0.0462, -1.6298, -1.6399, 0.0167], [0.6968, -0.1793, 0.5698, 0.8613, -1.8561], [0.7462, -0.1504, 0.0779, 2.0298, 1.2302], [-2.1399, -2.0118, -0.5827, 0.1486, 2.2127], [-0.2679, -0.5797, 0.5805, -0.4121, 0.5089], [1.3931, -0.8098, -0.3136, 0.6375, 0.6038], [1.8502, 0.0844, 0.7034, -0.1410, 2.5020]], requires_grad=True)

In [18]:

target

Out[18]:

tensor([[2.1635, 2.8280, -1.2495, 0.3782, 0.7208], [0.4892, 0.4965, -0.1423, 0.4918, -1.0321], [-1.4843, 0.4281, 0.6281, -1.4526, -1.8356], [0.7769, 2.5248, -0.4420, 0.4313, -1.0156], [-1.4197, -1.0438, 1.0570, 0.3100, 0.6264], [-1.3284, -0.9601, 0.0358, 0.5170, 1.5762], [0.2165, -1.0205, 0.2125, 0.4595, 0.9997], [0.4572, -0.3321, 0.3248, -0.4419, -0.0550], [-1.6006, 0.4164, -0.5147, -1.0651, -1.8708], [-1.9251, 0.9669, -0.9007, -0.4605, -0.1377]])

In [19]:

```python
def dloss_fn(t_p, t_c):
    dsq_diffs = 2 * (t_p - t_c)
    return dsq_diffs
```

In [20]:

```python
def model(t_u, w, b):
    return w * t_u + b
```

In [21]:

```python
def dmodel_dw(t_u, w, b):
    return t_u
```

In [22]:

```python
def dmodel_db(t_u, w, b):
    return 1.0
```

In [23]:

```python
def grad_fn(t_u, t_c, t_p, w, b):
    dloss_dw = dloss_fn(t_p, t_c) * dmodel_dw(t_u, w, b)
    dloss_db = dloss_fn(t_p, t_c) * dmodel_db(t_u, w, b)
    return torch.stack([dloss_dw.mean(), dloss_db.mean()])
```

参数配置包括输入数据、偏置值设定、学习率以及模型训练的迭代次数。通过对这些参数的估计来计算方程的值。

In [24]:

```
params = torch.tensor([1.0, 0.0])

nepochs = 10

learning_rate = 0.005

for epoch in range(nepochs):
    # 前向传播
    w, b = params
    t_p = model(t_u, w, b)

    loss = loss_fn(t_p, t_c)
    print('Epoch %d, Loss %f' % (epoch, float(loss)))
    # 反向传播
    grad = grad_fn(t_u, t_c, t_p, w, b)

    print('Params:', params)
    print('Grad:', grad)

    params = params - learning_rate * grad
params
Epoch 0, Loss 5259.733398
Params: tensor([1., 0.])
Grad: tensor([19936.2676,    143.4667])
Epoch 1, Loss 186035504.000000
Params: tensor([-98.6813,   -0.7173])
Grad: tensor([-3752242.2500,   -27117.4902])
Epoch 2, Loss 6590070521856.000000
Params: tensor([18662.5293,    134.8701])
Grad: tensor([7.0622e+08, 5.1037e+06])...................
```

上边是代码运行的初始结果。Epoch 代表了迭代的轮次数，每一轮迭代都会通过预设的损失函数计算出一个损失值。我们的目标是使损失函数最小化。参数向量（params）包含了需要调整的系数 w 和偏置值 b。通过 grad 函数，可以计算出向下一轮迭代传递的反馈值。这只是一个示例。选择合适的迭代轮次数是一项需要多次迭代的任务，具体取决于输入数据、输出数据，以及损失函数和优化函数的选择。

在训练模型时,降低学习率能够确保相关值正确地传递至梯度,优化参数更新,进而促使模型在数次迭代内实现收敛。

In [25]:

```
params = torch.tensor([1.0, 0.0])

nepochs = 10

learning_rate = 0.1

for epoch in range(nepochs):
    # 前向传播
    w, b = params
    t_p = model(t_u, w, b)

    loss = loss_fn(t_p, t_c)
    print('Epoch %d, Loss %f' % (epoch, float(loss)))
    # 反向传播
    grad = grad_fn(t_u, t_c, t_p, w, b)

    print('Params:', params)
    print('Grad:', grad)

    params = params - learning_rate * grad

params
```

```
Epoch 0, Loss 5259.733398
Params: tensor([1., 0.])
Grad: tensor([19936.2676,    143.4667])
Epoch 1, Loss 75167318016.000000
Params: tensor([-1992.6268,    -14.3467])
Grad: tensor([-75423624.0000,  -545075.6875])
Epoch 2, Loss 10758612701014917112.000000
Params: tensor([7540370.0000,  54493.2227])
Grad: tensor([2.8535e+11, 2.0621e+09])...............
```

如果学习率降低的幅度较小,权重的更新过程将相应地减缓,这表示需要增加迭代次数以确保模型达到稳定状态。

In [26]:

```
t_un = 0.1 * t_u
```

In [27]:
```python
params = torch.tensor([1.0, 0.0])

nepochs = 10

learning_rate = 0.05

for epoch in range(nepochs):
    # 前向传播
    w, b = params
    t_p = model(t_un, w, b)

    loss = loss_fn(t_p, t_c)
    print('Epoch %d, Loss %f' % (epoch, float(loss)))
    # 反向传播
    grad = grad_fn(t_un, t_c, t_p, w, b)

    print('Params:', params)
    print('Grad:', grad)

    params = params - learning_rate * grad

params
```

以下是结果：
```
Epoch 0, Loss 2642.455322
Params: tensor([1., 0.])
Grad: tensor([-1412.0094,  -102.6533])
Epoch 1, Loss 855426.562500
Params: tensor([71.6005,  5.1327])
Grad: tensor([25443.8555,  1838.2997])
Epoch 2, Loss 277741792.000000
Params: tensor([-1200.5923,   -86.7823])
Grad: tensor([-458472.5938, -33135.7656])...................
```

如果增加迭代的次数，那么损失函数和参数张量将会有所变化。可以通过以下脚本观察这些变化，其中我们打印出损失值，以便找到与迭代次数相对应的最小损失。然后，可以从模型中提取最佳参数。

In [28]:
```python
params = torch.tensor([1.0, 0.0])
```

```python
nepochs = 50
learning_rate = 1e-2
for epoch in range(nepochs):
    # 前向传播
    w, b = params
    t_p = model(t_un, w, b)

    loss = loss_fn(t_p, t_c)
    print('Epoch %d, Loss %f' % (epoch, float(loss)))
    # 反向传播
    grad = grad_fn(t_un, t_c, t_p, w, b)

    params = params - learning_rate * grad

params
```

以下是结果：

```
Epoch 0, Loss 2642.455322
Epoch 1, Loss 20719.347656
Epoch 2, Loss 162827.593750
Epoch 3, Loss 1279985.125000
Epoch 4, Loss 10062318.000000
Epoch 5, Loss 79103048.000000
Epoch 6, Loss 621853952.000000
Epoch 7, Loss 4888591872.000000
Epoch 8, Loss 38430781440.000000
```

以下是最终几轮迭代后的损失值。梯度爆炸问题是指由于初始参数设置不当或学习率不正确，或者两者共同作用导致的问题。为了解决这个问题，需要在初始化时使用裁剪或者在梯度计算中引入梯度裁剪。

```
Epoch 37, Loss 35813325661112375999719958590175313 92.000000
Epoch 38, Loss inf
Epoch 39, Loss inf
Epoch 40, Loss inf
Epoch 41, Loss inf
Epoch 42, Loss inf
Epoch 43, Loss inf
Epoch 44, Loss inf
```

Epoch 45, Loss inf
Epoch 46, Loss inf
Epoch 47, Loss inf
Epoch 48, Loss inf
Epoch 49, Loss inf

Out[28]:

tensor([-9.0577e+22, -6.5463e+21])

若要对该模型的参数进行微调，则需重新构建模型及损失函数，再将其应用于相同的示例。

In[29]:

```
def model(t_u, w, b):
    return w * t_u + b
```

In[30]:

```
def loss_fn(t_p, t_c):
    sq_diffs = (t_p - t_c)**2
    return sq_diffs.mean()
```

先设置参数向量。在完成（每一轮）训练过程后，应将 grad 函数重置为 None。

In[31]:

```
params = torch.tensor([1.0, 0.0], requires_grad=True)
loss = loss_fn(model(t_u, *params), t_c)
```

In[32]:

```
params.grad is None
```

Out[32]:

True

秘籍 3-3　模型微调

问题

如何通过应用优化函数来优化损失函数并找到梯度？

解决方案

要使用优化函数来优化损失函数并找到梯度，则应采用一种称为反向传播的方法。该方法在 PyTorch 中通过调用 backward() 函数来实现。通过 backward() 函数，可以计算损失函数对模型参数的梯度，进而利用优化算法调整模型参数以最小化损失函数。

编程实战

接下来，将通过一个实例展示如何使用 backward() 函数计算给定参数的函数的梯度。在接下来的内容中，将采用一组新的超参数对模型进行重新训练。

In[33]:
```
loss.backward()
```
In[34]:
```
params.grad
```
Out[34]:
```
tensor([19936.2676, 143.4667])
```

在当前的会话中，为了避免来自其他会话的错误值累积并产生混淆，必须**重置参数网格**。不进行重置可能会导致数据混淆和模型训练不稳定。因此，为了确保实验的准确性和可靠性，重置参数网格是必要的操作。

In[35]:
```
if params.grad is not None:
    params.grad.zero_()
```
In[36]:
```
def model(t_u, w, b):
    return w * t_u + b
```
In[37]:
```
def loss_fn(t_p, t_c):
    sq_diffs = (t_p - t_c)**2
    return sq_diffs.mean()
```

在完成对模型和损失函数的重新定义后，将开始对模型进行重新训练。

In[38]:

```
params = torch.tensor([1.0, 0.0], requires_grad=True)

nepochs = 5000

learning_rate = 1e-2
```

In[39]:

```
for epoch in range(nepochs):
    # 前向传播
    t_p = model(t_un, *params)
    loss = loss_fn(t_p, t_c)

    print('Epoch %d, Loss %f' % (epoch, float(loss)))
    # 反向传播
    if params.grad is not None:
        params.grad.zero_()

    loss.backward()

    #params.grad.clamp_(-1.0, 1.0)
    #print(params, params.grad)

    params = (params - learning_rate * params.grad).detach().requires_grad_()

params
```

Out[39]:

tensor([nan, nan], requires_grad=True)

秘籍 3-4　优化函数选择

问题

如何使用秘籍 3-3 中的 backward( ) 函数来优化梯度？

解决方案

在 PyTorch 中,优化函数既有内置的,也有需要用户自行创建的。内置的优化函数可以直接使用,而需要用户自行创建的优化函数则需要根据具体需求进行编写。

编程实战

让我们来看下面的示例代码。

In[40]:
```
import torch.optim as optim

dir(optim)
```

Out[40]:
```
['ASGD', 'Adadelta', 'Adagrad', 'Adam', 'AdamW', 'Adamax', 'LBFGS',
'NAdam', 'Optimizer', 'RAdam', 'RMSprop', 'Rprop', 'SGD', 'SparseAdam',
'__builtins__', '__cached__', '__doc__', '__file__', '__loader__',
'__name__', '__package__', '__path__', '__spec__', '_functional',
'_multi_tensor', 'lr_scheduler', 'swa_utils']
```

每一种优化方法在解决问题时都是独特的,后续将逐一深入探讨。

Adam 优化器是基于一阶随机目标函数的梯度优化方法,其设计理念是对低阶矩进行自适应估计,从而确保在大型数据集上具备高效的计算性能。在使用 PyTorch 的优化模块 torch.optim 时,需要在代码中构建一个优化器对象,该对象将负责维护参数的当前状态,并根据计算出的梯度、矩和学习率进行参数更新。在创建优化器实例时,需要为其提供一个可迭代对象,其中包含要优化的参数。同时,需要确保这些参数都是要优化的变量。此外,还可以根据需要为优化器指定特定的参数选项,例如学习率、权重衰减、冲量等。

SGD(随机梯度下降)是一种有效的优化器,适用于处理大型数据集。相较于其他优化器,SGD 不需要人工干预来调整学习率,而是通过算法内部机制进行自动调整。

In[41]:
```
params = torch.tensor([1.0, 0.0], requires_grad=True)
learning_rate = 1e-5
optimizer = optim.SGD([params], lr=learning_rate)
```
In[42]:
```
t_p = model(t_u, *params)
loss = loss_fn(t_p, t_c)
loss.backward()
optimizer.step()
params
```
Out[42]:
```
tensor([ 0.8006, -0.0014], requires_grad=True)
```
In[43]:
```
params = torch.tensor([1.0, 0.0], requires_grad=True)
learning_rate = 1e-2
optimizer = optim.SGD([params], lr=learning_rate)
t_p = model(t_un, *params)
loss = loss_fn(t_p, t_c)
optimizer.zero_grad()
loss.backward()
optimizer.step()
params
```
Out[43]:
```
tensor([15.1201, 1.0265], requires_grad=True)
```

接下来，将再次利用模型和损失函数，并对其进行优化处理。

In[44]:
```
def model(t_u, w, b):
    return w * t_u + b
```

In [45]:
```
def loss_fn(t_p, t_c):
    sq_diffs = (t_p - t_c)**2
    return sq_diffs.mean()
```
In [46]:
```
params = torch.tensor([1.0, 0.0], requires_grad=True)

nepochs = 5000
learning_rate = 1e-2

optimizer = optim.SGD([params], lr=learning_rate)
```
In [47]:
```
for epoch in range(nepochs):
    # 前向传播
    t_p = model(t_un, *params)
    loss = loss_fn(t_p, t_c)
    print('Epoch %d, Loss %f' % (epoch, float(loss)))
    # 反向传播
    optimizer.zero_grad()
    loss.backward()
    optimizer.step()

t_p = model(t_un, *params)

params
```

在探讨损失函数时，应当关注其梯度。利用优化库可以尝试寻找损失函数的最佳值。

在提供的例子中，存在两个自定义函数和一个损失函数。在此例中，使用了两个较小的张量值。与前述章节不同的是，本次我们利用优化器来寻找最小值。

在以下示例中，将采用 Adam 作为优化器。

In [48]:
```
def model(t_u, w, b):
    return w * t_u + b
```
In [49]:
```
def loss_fn(t_p, t_c):
    sq_diffs = (t_p - t_c)**2
    return sq_diffs.mean()
```

In[50]:

```python
params = torch.tensor([1.0, 0.0], requires_grad=True)

nepochs = 5000
learning_rate = 1e-1

optimizer = optim.Adam([params], lr=learning_rate)
```

In[51]:

```python
for epoch in range(nepochs):
    # 前向传播
    t_p = model(t_u, *params)
    loss = loss_fn(t_p, t_c)

    print('Epoch %d, Loss %f' % (epoch, float(loss)))
    # 反向传播
    optimizer.zero_grad()
    loss.backward()
    optimizer.step()

t_p = model(t_u, *params)

params
```

```
Epoch 0, Loss 5259.733398
Epoch 1, Loss 3443.706543
Epoch 2, Loss 2025.263306
Epoch 3, Loss 1002.202881
Epoch 4, Loss 357.638672
Epoch 5, Loss 53.362339
Epoch 6, Loss 24.627544
Epoch 7, Loss 181.475266
Epoch 8, Loss 421.412598
Epoch 9, Loss 651.219666
Epoch 10, Loss 806.726135
...
Epoch 4993, Loss 0.167906
Epoch 4994, Loss 0.167906
Epoch 4995, Loss 0.167905
Epoch 4996, Loss 0.167904
```

```
Epoch 4997, Loss 0.167903
Epoch 4998, Loss 0.167903
Epoch 4999, Loss 0.167903
```

Out[51]:

```
tensor([ 0.2879, 25.6386], requires_grad=True)
```

在上述代码片段中,我们对参数进行了优化,并计算了预测张量。为了直观地展示结果,采用回归曲线图来展示预测结果。

接下来,将对实际张量和预测张量的样本数据进行可视化,可视化结果如图3-1所示。

In[52]:

```
from matplotlib import pyplot as plt
%matplotlib inline
plt.plot(0.1 * t_u.numpy(), t_p.detach().numpy())
plt.plot(0.1 * t_u.numpy(), t_c.numpy(), 'o')
```

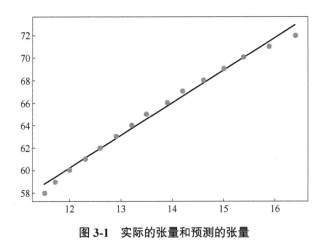

图 3-1　实际的张量和预测的张量

秘籍 3-5　进一步优化函数

问题

如何使用随机样本在训练集上优化模型并在验证集上进行测试?

解决方案

以下是进一步优化的过程。

编程实战

在下边的示例中,首先设定样本数量,接着利用 shuffled_indices 变量对数据集实施随机划分,将全体记录的 20% 作为验证样本。需要注意的是,这里的随机抽样针对的是所有数据记录。拆分训练集与验证集的目标在于:在训练集上构建模型,然后在验证集上进行预测,以评估模型准确性。

In[53]:
```
n_samples = t_u.shape[0]
n_val = int(0.2 * n_samples)

shuffled_indices = torch.randperm(n_samples)

train_indices = shuffled_indices[:-n_val]
val_indices = shuffled_indices[-n_val:]

train_indices, val_indices
```
Out[53]:
(tensor([14, 13, 7, 6, 11, 0, 10, 3, 4, 5, 12, 1]), tensor([8, 2, 9]))

In[54]:
```
t_u_train = t_u[train_indices]
t_c_train = t_c[train_indices]

t_u_val = t_u[val_indices]
t_c_val = t_c[val_indices]
```
In[55]:
```
def model(t_u, w, b):
    return w * t_u + b
```
In[56]:
```
def loss_fn(t_p, t_c):
    sq_diffs = (t_p - t_c)**2
    return sq_diffs.mean()
```

In [57]:
```
params = torch.tensor([1.0, 0.0], requires_grad=True)

nepochs = 5000
learning_rate = 1e-2

optimizer = optim.SGD([params], lr=learning_rate)

t_un_train = 0.1 * t_u_train
t_un_val = 0.1 * t_u_val
```

接下来，展开训练与验证过程：在前向传播阶段，首先选取训练集输入数据与参数相乘，随后进行预测并计算损失函数。紧接着，在代码的第三行（验证模型阶段），调用相同模型，输入验证数据进行预测，并计算验证集的损失函数。在反向传播阶段，计算训练集损失函数的梯度，并借助优化器更新参数。

In [58]:
```
for epoch in range(nepochs):
    # 前向传播
    t_p_train = model(t_un_train, *params)
    loss_train = loss_fn(t_p_train, t_c_train)
    t_p_val = model(t_un_val, *params)
    loss_val = loss_fn(t_p_val, t_c_val)

    print('Epoch %d, Training loss %f, Validation loss %f' % (epoch,
    float(loss_train),
                                                              float
                                                              (loss_val)))

    # 反向传播
    optimizer.zero_grad()
    loss_train.backward()
    optimizer.step()

t_p = model(t_un, *params)

params

Epoch 0, Training loss 2652.548340, Validation loss 2602.083252
Epoch 1, Training loss 21507.599609, Validation loss 19826.755859
Epoch 2, Training loss 174840.953125, Validation loss 164931.500000
Epoch 3, Training loss 1421780.000000, Validation loss 1330612.750000
```

Epoch 4, Training loss 11562152.000000, Validation loss 10851068.000000
Epoch 5, Training loss 94025800.000000, Validation loss 88156856.000000
Epoch 6, Training loss 764637760.000000, Validation loss 717156288.000000
Epoch 7, Training loss 6218196992.000000, Validation loss 5831366144.000000
Epoch 8, Training loss 50567704576.000000, Validation loss 47423901696.000000

以下是最后 11 个训练轮次及结果：

Epoch 4989, Training loss nan, Validation loss nan
Epoch 4990, Training loss nan, Validation loss nan
Epoch 4991, Training loss nan, Validation loss nan
Epoch 4992, Training loss nan, Validation loss nan
Epoch 4993, Training loss nan, Validation loss nan
Epoch 4994, Training loss nan, Validation loss nan
Epoch 4995, Training loss nan, Validation loss nan
Epoch 4996, Training loss nan, Validation loss nan
Epoch 4997, Training loss nan, Validation loss nan
Epoch 4998, Training loss nan, Validation loss nan
Epoch 4999, Training loss nan, Validation loss nan

Out[58]:

tensor([nan, nan], requires_grad=True)

在前边的操作中，已将梯度设置为启用。在后续的设置中，可通过应用 torch.no_grad( ) 函数来禁用梯度计算。其他语句保持不变。禁用梯度计算在确认不会调用 Tensor.backward( ) 时进行推理非常有用，可以降低那些需要设置 requires_grad=True 的张量的内存消耗。

In[59]:

```
for epoch in range(nepochs):
    # 前向传播
    t_p_train = model(t_un_train, *params)
    loss_train = loss_fn(t_p_train, t_c_train)

    with torch.no_grad():
        t_p_val = model(t_un_val, *params)
        loss_val = loss_fn(t_p_val, t_c_val)

    print('Epoch %d, Training loss %f, Validation loss %f' % (epoch,
        float(loss_train),
                                                              float
                                                              (loss_val)))
```

```
# 反向传播
optimizer.zero_grad()
loss_train.backward()
optimizer.step()

params
Epoch 0, Training loss nan, Validation loss nan
Epoch 1, Training loss nan, Validation loss nan
Epoch 2, Training loss nan, Validation loss nan
Epoch 3, Training loss nan, Validation loss nan
Epoch 4, Training loss nan, Validation loss nan
Epoch 5, Training loss nan, Validation loss nan
Epoch 6, Training loss nan, Validation loss nan
Epoch 7, Training loss nan, Validation loss nan
Epoch 8, Training loss nan, Validation loss nan
Epoch 9, Training loss nan, Validation loss nan
Epoch 10, Training loss nan, Validation loss nan............
```

Out[59]:

tensor([nan, nan], requires_grad=True)

最后几轮迭代的结果参见本书代码，最终的参数是 5.44 和 −18.012。

秘籍 3-6　实现卷积神经网络

问题

如何使用 PyTorch 实现卷积神经网络？

解决方案

在 torchvision 库中，内置了众多可供使用的数据集。在此，以 MNIST 数据集为例，尝试构建一个卷积神经网络模型。

编程实战

以下是一个实例。首先,需设定超参数;其次,设计网络结构;最后,训练模型并进行预测。

In [60]:

```
# CNN:卷积神经网络
```

In [61]:

```
import torch
import torch.nn as nn
from torch.autograd import Variable
import torch.utils.data as Data
import torchvision
import matplotlib.pyplot as plt
%matplotlib inline
```

In [62]:

```
torch.manual_seed(1)   # 确保可复现
```

在上述代码段中,我们导入了所需的库以构建用于 digits 数据集(见图 3-2)的卷积神经网络模型。在深度学习领域,MNIST 数字数据集作为计算机视觉和图像处理方面的常用数据集,具有极高的知名度。

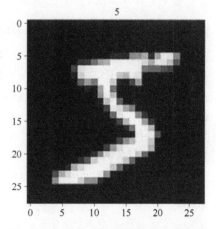

图 3-2　MNIST 数字数据集中图像示例

In [63]:

```
# 超参数设置
EPOCH = 1
# 需要对输入数据进行n轮训练,为了节省时间,这里只训练一个周期
BATCH_SIZE = 50
# 每个轮次使用50个样本进行训练迭代
LR = 0.001
# 学习率
DOWNLOAD_MNIST = True
# 如果已下载数据,请设置为False
```

In [64]:
```python
#Mnist手写数字数据集
train_data = torchvision.datasets.MNIST(
    root='./mnist/',
    train=True,
    # 表示这是训练数据
    transform=torchvision.transforms.ToTensor(),
    # 形状为(Color x Height x Width)的torch.FloatTensor
    # 在[0.0, 1.0]范围内进行归一化
    download=DOWNLOAD_MNIST,
    # 如果没有数据集，则需要进行下载
)
```

In [65]:
```python
# 绘制一个数据实例
print(train_data.train_data.size())   # 训练数据集数据尺寸（60000 , 28 , 28）
print(train_data.train_labels.size())   # 训练数据集标签尺寸（60000）
plt.imshow(train_data.train_data[0].numpy(), cmap='gray')
plt.title('%i' % train_data.train_labels[0])
plt.show()
torch.Size([60000, 28, 28])
torch.Size([60000])
```

接下来，将使用数据加载函数（DataLoader）导入数据集。

In [66]:
```python
# 数据加载器用于在训练中返回小批量数据，图像批次的形状将为
#(50, 1, 28, 28)
train_loader = Data.DataLoader(dataset=train_data, batch_size=BATCH_SIZE, shuffle=True)
```

In [67]:
```python
# 将测试数据的数据类型转换为Variable，选择2000个样本以加快测试速度
test_data = torchvision.datasets.MNIST(root='./mnist/', train=False)
test_x = Variable(torch.unsqueeze(test_data.test_data, dim=1)).type(torch.FloatTensor)[:2000]/255.
# 数据形状从(2000, 28, 28)变为(2000, 1, 28, 28)，值在范围(0,1)内
```

```
test_y = test_data.test_labels[:2000]
```

在卷积神经网络架构中，输入图像会被转换为一个特征集，其特征数量等于图像的通道数乘以图像的高度和宽度。由于数据集的维度过高，无法直接应用于建模以预测输出。若输出层包括车、卡车、面包车和自行车等类别，输入自行车图像时，输入图像中应包含卷积神经网络模型可利用的特征并正确预测结果。卷积层通常搭配池化层，如最大池化和平均池化。通过不断叠加池化层和卷积层，可以将维度降低至适用于简单全连接神经网络预测正确类别的程度。如图 3-3 所示。

In [68]:
```python
class CNN(nn.Module):
    def __init__(self):
        super(CNN, self).__init__()
        self.conv1 = nn.Sequential(         # 输入形状 (1, 28, 28)
            nn.Conv2d(
                in_channels=1,              # 输入高度
                out_channels=16,            # 等同 n_filters, 卷积核个数
                kernel_size=5,              # 卷积核大小
                stride=1,                   # 卷积核移动步长
                padding=2,
                # 如果希望在经过 Conv2d 后图像的宽度和长度保持不变,
                # 当步长(stride)为 1 时, 填充(padding)应为 (kernel_size-1)/2
            ),                              # 输出形状为 (16, 28, 28)
            nn.ReLU(),                      # 激活函数
            nn.MaxPool2d(kernel_size=2),
            # 在 2x2 区域内选择最大值, 输出形状为 (16, 14, 14)
        )
        self.conv2 = nn.Sequential(         # 输入形状 (1, 28, 28)
            nn.Conv2d(16, 32, 5, 1, 2),     # 输出形状 (32, 14, 14)
            nn.ReLU(),                      # 激活函数
            nn.MaxPool2d(2),                # 输出形状 (32, 7, 7)
        )
```

```python
        self.out = nn.Linear(32 * 7 * 7, 10)    # 全连接层，输出 10 个类别
    def forward(self, x):
        x = self.conv1(x)
        x = self.conv2(x)
        x = x.view(x.size(0), -1)
        # 将 conv2 的输出展平为 (batch_size, 32 * 7 * 7)
        output = self.out(x)
        return output, x    # 返回 x 以供可视化
```
In [69]:
```python
cnn = CNN()
print(cnn)    # 网络架构
CNN(
  (conv1): Sequential(
    (0): Conv2d(1, 16, kernel_size=(5, 5), stride=(1, 1), padding=(2, 2))
    (1): ReLU()
    (2): MaxPool2d(kernel_size=2, stride=2, padding=0, dilation=1,
        ceil_mode=False)
  )
  (conv2): Sequential(
    (0): Conv2d(16, 32, kernel_size=(5, 5), stride=(1, 1), padding=(2, 2))
    (1): ReLU()
    (2): MaxPool2d(kernel_size=2, stride=2, padding=0, dilation=1,
        ceil_mode=False)
  )
  (out): Linear(in_features=1568, out_features=10, bias=True)
)
```
In [70]:
```python
optimizer = torch.optim.Adam(cnn.parameters(), lr=LR)    # 优化所有 cnn 参数
loss_func = nn.CrossEntropyLoss()                        # 目标标签不是独热标签
```
In [71]:
```python
import sklearn
import warnings
warnings.filterwarnings("ignore", category=FutureWarning)
import warnings
warnings.filterwarnings("ignore")
```

In[72]:

```python
from matplotlib import cm
try: from sklearn.manifold import TSNE; HAS_SK = True
except: HAS_SK = False; print('Please install sklearn for layer visualization, if not there')
def plot_with_labels(lowDWeights, labels):
    plt.cla()
    X, Y = lowDWeights[:, 0], lowDWeights[:, 1]
    for x, y, s in zip(X, Y, labels):
        c = cm.rainbow(int(255 * s / 9)); plt.text(x, y, s,
        backgroundcolor=c, fontsize=9)
    plt.xlim(X.min(), X.max()); plt.ylim(Y.min(), Y.max()); plt.title('Visualize last layer');
    plt.show();
    #plt.pause(0.01)
plt.ion()
# 训练和测试
for epoch in range(EPOCH):
    for step, (x, y) in enumerate(train_loader):
        # 提供批量数据，并在迭代训练加载器（train_loader）的过程中对 x 进行标
            准化处理
        b_x = Variable(x)       # 批量输入数据 x
        b_y = Variable(y)       # 批量输出标签 y
        output = cnn(b_x)[0]                    # cnn 输出
        loss = loss_func(output, b_y)           # 交叉熵损失
        optimizer.zero_grad()                   # 清除此训练步骤的梯度
        loss.backward()                         # 反向传播，计算梯度
        optimizer.step()                        # 通过梯度下降执行一步参数更新
        if step % 100 == 0:
            test_output, last_layer = cnn(test_x)
            pred_y = torch.max(test_output, 1)[1].data.squeeze()
            accuracy = (pred_y == test_y).sum().item() / float(test_y.size(0))
            print('Epoch: ', epoch, '| train loss: %.4f' % loss.data,
                  '| test accuracy: %.2f' % accuracy)
            if HAS_SK:
```

```
# 训练后的扁平化层可视化 (T-SNE)
tsne = TSNE(perplexity=30, n_components=2, init='pca',
n_iter=5000)
plot_only = 500
low_dim_embs = tsne.fit_transform(last_layer.data.numpy())
[:plot_only, :])
labels = test_y.numpy()[:plot_only]
plot_with_labels(low_dim_embs, labels)
```
plt.ioff()

在图 3-3 的第一张图中，如果观察数字 4，会发现它在整个图中出现的位置十分分散。理想情况下，所有的 4 应该彼此更接近。这是因为训练前模型测试准确率较低。

在第一个循环中，训练损失从 0.4369 降低到 0.1482，而测试准确率从 16% 提高到 94%。具有相同颜色的数字在图中被放置在十分相近的位置。

在接下来的轮次中，对于 MNIST 数字数据集，测试准确率提高到了 95%。

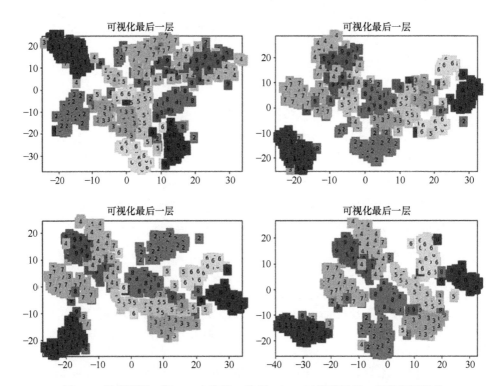

图 3-3　随着训练，每 100 个步长，使用 TSNE 对模型最后一层进行可视化

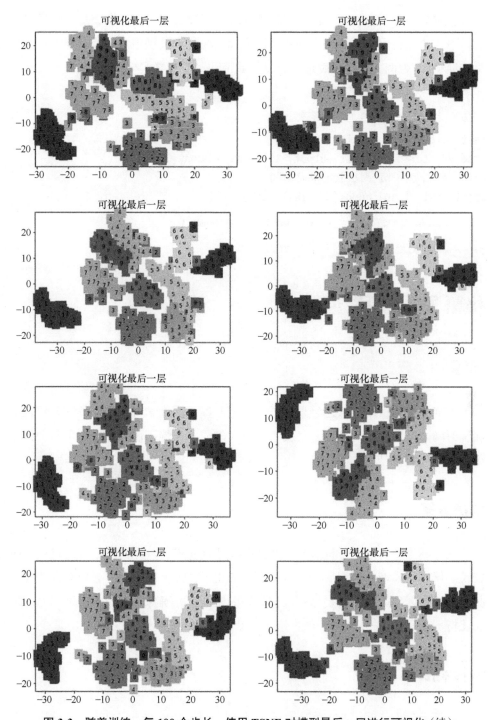

图 3-3 随着训练,每 100 个步长,使用 TSNE 对模型最后一层进行可视化(续)

在最后一步/轮次中，具有相同颜色的数字被放在一起。在成功训练模型之后，下一步是利用该模型进行预测。以下代码解释了模型预测过程。我们以测试集的前 10 个为例。代码最后输出预测的数字和对应的真实标签。

In [73]:

```
# 输出测试数据中的 10 个预测结果
test_output, _ = cnn(test_x[:10])
pred_y = torch.max(test_output, 1)[1].data.numpy().squeeze()
print(pred_y, 'prediction number')
print(test_y[:10].numpy(), 'real number')
[7 2 1 0 4 1 4 9 5 9] prediction number
[7 2 1 0 4 1 4 9 5 9] real number
```

秘籍 3-7　模型重载

问题

如何存储和重新加载已经训练过的模型？鉴于深度学习模型通常需要较长的训练时间，计算过程对公司来说会带来巨大的成本。能否使用新数据重新训练模型并存储该模型？

解决方案

在实际生产场景中，训练与预测往往无法同步进行，原因在于训练过程耗时较长。仅在完成训练过程后，方可将模型应用于预测。因此，有必要对训练与预测环节进行区分。我们需要储存已训练好的模型，并确保其在下一次训练完成之前保持不变。

编程实战

以下示例中，我们定义了 save( ) 函数，用于利用 Torch 神经网络模块保存模型；同时定义了 restore_net( ) 函数，用以恢复先前训练过的神经网络模型。

In [74]:
保存并重新加载

In [75]:
```python
import torch
from torch.autograd import Variable
import matplotlib.pyplot as plt
%matplotlib inline
torch.manual_seed(1)    # 为保证可复现
```

In [76]:
```python
# 样本数据
x = torch.unsqueeze(torch.linspace(-1, 1, 100), dim=1)  # x 数据格式为 tensor, 形状为(100,1)
y = x.pow(2) + 0.2*torch.rand(x.size())    # 噪声数据 y 数据格式为 tensor, 形状为(100,1)
x, y = Variable(x, requires_grad=False), Variable(y, requires_grad=False)
```

上述脚本中,我们引入了目标变量 Y 与独立变量 X,以此作为样本数据点来训练神经网络模型。紧接着,利用 save( ) 函数对所构建的模型进行保存。net1 代表已训练好的神经网络模型,其存储方式有两种:①保存整个神经网络模型及其全部权重和偏置;②仅保存模型参数的权重。在训练模型参数量较大时,建议仅存储权重参数;若参数量较小,可以考虑保存整个模型结构。

In [77]:
```python
def save():
    # 保存 net1
    net1 = torch.nn.Sequential(
        torch.nn.Linear(1, 10),
        torch.nn.ReLU(),
        torch.nn.Linear(10, 1)
    )
    optimizer = torch.optim.SGD(net1.parameters(), lr=0.5)
    loss_func = torch.nn.MSELoss()

    for t in range(100):
        prediction = net1(x)
        loss = loss_func(prediction, y)
```

```
        optimizer.zero_grad()
        loss.backward()
        optimizer.step()
    # 绘制结果
    plt.figure(1, figsize=(10, 3))
    plt.subplot(131)
    plt.title('Net1')
    plt.scatter(x.data.numpy(), y.data.numpy())
    plt.plot(x.data.numpy(), prediction.data.numpy(), 'r-', lw=5)
    # 两种保存网络的方法
    torch.save(net1, 'net.pkl')  # 保存整个网络
    torch.save(net1.state_dict(), 'net_params.pkl')  # 仅保存网络参数
```

我们可以利用 torch.load( ) 函数，将预先训练好的神经网络模型重新加载至现有的 PyTorch 环境中。为了对 net1 对象进行测试并实现预测，需要先加载 net1 保存的模型，然后将其命名为 net2。借助 net2 对象，便可对目标变量进行预测。代码最后一行的 prediction.data.numpy( ) 用于展示预测结果。

In [78]:
```
def restore_net():
    # 将net1网络加载恢复到net2中
    net2 = torch.load('net.pkl')
    prediction = net2(x)
    # 绘制结果
    plt.subplot(132)
    plt.title('Net2')
    plt.scatter(x.data.numpy(), y.data.numpy())
    plt.plot(x.data.numpy(), prediction.data.numpy(), 'r-', lw=5)
```

加载 pickle 文件格式的整个神经网络是一个相对较慢的过程；然而，在仅针对新数据集进行预测的情况下，只需加载模型参数，而无须加载整个网络。

In [79]:
```
def restore_params():
# 只将net1中的参数恢复到net3中
net3 = torch.nn.Sequential(
    torch.nn.Linear(1, 10),
```

```
    torch.nn.ReLU(),
    torch.nn.Linear(10, 1)
)
    # 将 net1 的参数复制到 net3 中
    net3.load_state_dict(torch.load('net_params.pkl'))
    prediction = net3(x)
    # 绘制结果
    plt.subplot(133)
    plt.title('Net3')
    plt.scatter(x.data.numpy(), y.data.numpy())
    plt.plot(x.data.numpy(), prediction.data.numpy(), 'r-', lw=5)
    plt.show()
```

In [80]:
```
# 保存 net1
save()
# 恢复整个网络(可能会变慢)
restore_net()
# 仅恢复网络参数
restore_params()
```

  模型重用。restore_params( ) 函数确保已训练的参数可被模型重新使用。为恢复模型，可调用 load_state_dict( ) 函数导入模型参数。如图 3-4 所示，三个模型是相同的，因为 net2 和 net3 为 net1 的副本。

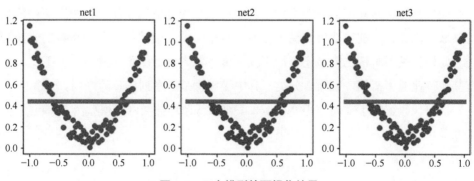

图 3-4　三个模型的可视化结果

秘籍 3-8　实现循环神经网络

问题

如何使用 MNIST 数据集设置一个循环神经网络呢？

解决方案

循环神经网络被认为是记忆网络的一种。在本例中，我们采用每个 epoch 包含 1 个完整周期，批处理大小为每次 64 个样本，以建立输入与输出之间的关联。通过使用循环神经网络模型，可以预测图像中所包含的数字。图 3-5 所示为 MNIST 数据集示例。

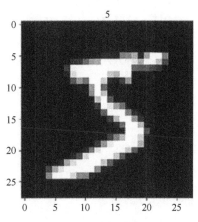

图 3-5　MNIST 数据集示例

编程实战

以下是一个实例。循环神经网络在输入层接收一系列向量，输出层生成相应的一系列向量。信息序列通过循环层中的内部状态进行传递处理。在某些情况下，输出值对过去历史值具有较长的依赖关系，如长短时记忆（LSTM）模型，它是循环神经网络模型的一个变种。长短时记忆模型适用于任何按顺序消耗信息的领域，例如，当前股票价格受历史股票价格影响，依赖关系可能是短期或长期的。此外，长短时记忆模型还可用于预测文本输入向量的长期和短期上下文。还有其他行业应用案例，如噪声分类，其中噪声也是一系列信息的序列。

以下代码解释了如何使用 PyTorch 模块执行循环神经网络模型。

In[81]:

```
# 循环神经网络
```

In[82]:

```
import torch
```

```python
from torch import nn
from torch.autograd import Variable
import torchvision.datasets as dsets
import torchvision.transforms as transforms
import matplotlib.pyplot as plt
%matplotlib inline
```

In[83]:

```python
torch.manual_seed(1)        # 为了可复现
```

In[84]:

```python
# 超参数设置
EPOCH = 1                   # 将训练数据训练n个轮次,为了节省时间,只训练1个周期
BATCH_SIZE = 64
TIME_STEP = 28              # 循环神经网络时间步长,即图像高度
INPUT_SIZE = 28             # 循环神经网络输入大小,即图像宽度
LR = 0.01                   # 学习率
DOWNLOAD_MNIST = True       # 如果尚未下载数据,则设置为True
```

循环神经网络模型包含若干超参数,如迭代次数(EPOCH)、批处理大小(BATCH_SIZE,依据单台计算机可用内存而定)、时间步长(TIME_STEP,用于保留信息序列)、输入大小(INPUT_SIZE,体现输入向量规模)以及学习率(LR)。这些参数值的选择具有针对性,未必适用于其他应用场景。超参数值的优化选择是一个迭代过程,可通过调整多个参数并进行比较,以确定哪些参数对模型具有积极作用,或者对模型进行并行训练,以判断哪个模型表现更为优异。

In[85]:

```python
#Mnist手写数字数据集
train_data = dsets.MNIST(
    root='./mnist/',
    train=True,                         # 这是训练数据
    transform=transforms.ToTensor(),    # 将PIL.Image 或 numpy.ndarray 转换为
                                        # 形状为(C x H x W)的 torch.FloatTensor,
                                        # 并在[0.0, 1.0]范围内进行归一化
    download=DOWNLOAD_MNIST,            # 如果没有数据集,则需要进行下载
)
```

In [86]:
```
# 绘制一个数据实例
print(train_data.train_data.size())          # 数据形状为 (60000, 28, 28)
print(train_data.train_labels.size())        # 标签形状为 (60000)
plt.imshow(train_data.train_data[0].numpy(), cmap='gray')
plt.title('%i' % train_data.train_labels[0])
plt.show()
torch.Size([60000, 28, 28])
torch.Size([60000])
```

使用 dsets.MINIST( ) 函数可以将数据集加载到当前会话中。如果需要将数据集存储在本地，那么需要先下载数据集。

通过运行上述脚本，展示了示例图像数据集的图像样例。在训练深度学习模型时，需将整个训练数据集分为小批次，这有助于对模型的最终准确性进行平均。通过数据加载器函数 DataLoader( )，可加载训练数据并将其划分为小批次。对小批次进行 shuffle 操作，确保模型能捕捉到实际数据集中的所有变化。

In [87]:
```
# 数据加载器用于在训练中返回小批量数据
train_loader = torch.utils.data.DataLoader(dataset=train_data,
                                           batch_size=BATCH_SIZE,
                                           shuffle=True)
```

In [88]:
```
# 将测试数据转换为 Variable, 选择 2000 个样本以加速测试
test_data = dsets.MNIST(root='./mnist/', train=False, transform=transforms.ToTensor())
test_x = Variable(test_data.test_data, volatile=True).type(torch.FloatTensor)[:2000]/255.
# 数据形状为 (2000, 28, 28), 值在范围 (0, 1) 内
test_y = test_data.test_labels.numpy().squeeze()[:2000]  # 转换为 NumPy 数组
```

在上述脚本中，首先构建了训练数据集。接着，通过设置标志 train=False，获取相应的测试数据。随后，对测试数据集进行随机抽样（每次抽取 2000 个样本），并将抽样结果转换为张量以进行模型测试。测试数据集被转换为 Variable 类型，测试标签向量则转换为 NumPy 数组类型。

In [89]:
```python
class RNN(nn.Module):
    def __init__(self):
        super(RNN, self).__init__()
        self.rnn = nn.LSTM(              # 如果使用nn.RNN(),则几乎无法拟合
            input_size=INPUT_SIZE,
            hidden_size=64,              # 循环神经网络隐藏单元
            num_layers=1,                # 循环神经网络层数
            batch_first=True,            # 输入和输出将包含批处理大小维度为1的数据
        )
        self.out = nn.Linear(64, 10)

    def forward(self, x):
        # x 的形状为(batch, time_step, input_size)
        # r_out 的形状为 (batch, time_step, output_size)
        # h_n 的形状为 (n_layers, batch, hidden_size)
        # h_c 的形状为 (n_layers, batch, hidden_size)
        r_out, (h_n, h_c) = self.rnn(x, None)   # None表示零初始隐藏状态
        # 在最后一个时间步长选择输出 r_out
        out = self.out(r_out[:, -1, :])
        return out
```

在上述的循环神经网络类中，我们训练的是一个 LSTM 网络。LSTM 网络已被证实对于长时间序列记忆具有显著优势，从而有利于参数学习。相反，如果采用 nn.RNN（）模型，其学习效果几乎受限，因为循环神经网络的原始实现方式难以在长时间内保持或记住信息。在 LSTM 网络中，图像的宽度被视为输入大小，hidden_size 定义了隐藏层中的神经元数量，而 num_layers 则表示网络中的循环神经网络层数。

在循环神经网络模块中（包括 LSTM 模块在内），其输出向量尺寸为 64×10，这是因为输出层需要对数字进行分类，范围为 0~9。最后的 forward（）函数演示了在循环神经网络中如何实现前向传播。

以下脚本演示在循环神经网络类内如何实现 LSTM 模型的构建。在 LSTM 函数中，输入长度设定为 28，隐藏层神经元数量为 64，并将这 64 个隐藏神经元

传递至输出层的 10 个神经元。

In [90]:
```
rnn = RNN()
print(rnn)
RNN(
  (rnn): LSTM(28, 64, batch_first=True)
  (out): Linear(in_features=64, out_features=10, bias=True)
)
```

In [91]:
```
optimizer = torch.optim.Adam(rnn.parameters(), lr=LR)   # 优化所有循环神经网络参数

loss_func = nn.CrossEntropyLoss()                        # 目标标签不是独热标签
```

为了优化循环神经网络的所有参数，我们采用了 Adam 优化器。在优化过程中，学习率在其中发挥着重要作用。在此示例中，使用了交叉熵损失函数作为损失函数。为了获得最佳参数，需要进行多个周期（epoch）的训练。

在下列脚本中，展示了训练损失值与测试准确率的变化。经过一个训练周期后，测试准确率上升至 95%，同时训练损失值下降至 0.24。

In [92]:
```
# 训练和测试
for epoch in range(EPOCH):
    for step, (x, y) in enumerate(train_loader):    # 获取批次的数据
        b_x = Variable(x.view(-1, 28, 28))          # 将x重塑为(batch, time_step, input_size)

        b_y = Variable(y)                            # 标签数据批次 y
        output = rnn(b_x)                            # rnn 的输出
        loss = loss_func(output, b_y)                # 交叉熵损失
        optimizer.zero_grad()                        # 清除此训练步骤的梯度
        loss.backward()                              # 反向传播，计算梯度
        optimizer.step()                             # 通过梯度下降执行一步参数更新

        if step % 50 == 0:
```

```
        test_output = rnn(test_x)              # 输出数据的尺寸为(samples,
                                                 time_step, input_size)
        pred_y = torch.max(test_output, 1)[1].data.numpy().squeeze()
        accuracy = sum(pred_y == test_y) / float(test_y.size)
        print('Epoch: ', epoch, '| train loss: %.4f' % loss.data,
        '| test accuracy: %.2f' % accuracy)

Epoch:  0 | train loss: 2.3088 | test accuracy: 0.09
Epoch:  0 | train loss: 1.3125 | test accuracy: 0.59
Epoch:  0 | train loss: 0.8936 | test accuracy: 0.71
Epoch:  0 | train loss: 0.4285 | test accuracy: 0.83
Epoch:  0 | train loss: 0.2509 | test accuracy: 0.87
Epoch:  0 | train loss: 0.3429 | test accuracy: 0.90
Epoch:  0 | train loss: 0.3704 | test accuracy: 0.86
Epoch:  0 | train loss: 0.4593 | test accuracy: 0.91
Epoch:  0 | train loss: 0.0794 | test accuracy: 0.94
Epoch:  0 | train loss: 0.0768 | test accuracy: 0.93
Epoch:  0 | train loss: 0.1809 | test accuracy: 0.94
Epoch:  0 | train loss: 0.2297 | test accuracy: 0.94
Epoch:  0 | train loss: 0.2210 | test accuracy: 0.95
Epoch:  0 | train loss: 0.2509 | test accuracy: 0.94
Epoch:  0 | train loss: 0.0828 | test accuracy: 0.94
Epoch:  0 | train loss: 0.2879 | test accuracy: 0.95
Epoch:  0 | train loss: 0.0908 | test accuracy: 0.94
Epoch:  0 | train loss: 0.1554 | test accuracy: 0.94
Epoch:  0 | train loss: 0.1557 | test accuracy: 0.96
```

在完成模型训练之后，接下来便是运用所述的循环神经网络模型进行预测。随后，将把实际输出值与预测输出值进行对比，从而对模型性能进行评估。

In [93]:
```
# 输出测试数据的10个预测结果
test_output = rnn(test_x[:10].view(-1, 28, 28))
pred_y = torch.max(test_output, 1)[1].data.numpy().squeeze()
print(pred_y, 'prediction number')
print(test_y[:10], 'real number')

[7 2 1 0 4 1 4 9 5 9] prediction number
[7 2 1 0 4 1 4 9 5 9] real number
```

秘籍 3-9 实现用于回归问题的循环神经网络

问题

如何为基于回归问题设置一个循环神经网络？

解决方案

回归模型构建需依托于目标函数、特征集以及构建输入与输出关系的函数。在以下实例中，我们采用循环神经网络实现回归任务。尽管回归问题在处理具有明确线性关系的数据时效果颇佳，但在预测输入与输出之间的非线性关系时，问题复杂度显著提升。

编程实战

以下是一个实例，其中展示了输入和输出数据之间的非线性周期性模式。在先前的教程中，我们提供了一个关于循环神经网络的一般性示例，用于解决与分类相关的问题，即预测输入图像的类别。然而，在回归问题中，循环神经网络的架构会发生变化，因为其目标是预测实数值输出。在回归相关的问题中，输出层仅包含一个神经元。

In[94]:
循环神经网络回归器

In[95]:
```
import torch
from torch import nn
from torch.autograd import Variable
import numpy as np
import matplotlib.pyplot as plt
%matplotlib inline
```

In [96]:
torch.manual_seed(1)   # 为保证可复现

In [97]:
# 超参数设置
TIME_STEP = 10   # 循环神经网络时间步长
INPUT_SIZE = 1   # 循环神经网络输入大小
LR = 0.02   # 学习率

循环神经网络的时间步长（TIME_STEP）代表着过去 10 个数值对当前值的预测，然后进行滚动预测。

以下脚本用于展示样本序列，后续章节将使用 sin 函数逼近目标 cos 函数（见图 3-6）。

In [98]:
# 显示数据
steps = np.linspace(0, np.pi*2, 100, dtype=np.float32)
x_np = np.sin(steps)   # float32 转换为 torch FloatTensor
y_np = np.cos(steps)
plt.plot(steps, y_np, 'r-', label='target (cos)')
plt.plot(steps, x_np, 'b-', label='input (sin)')
plt.legend(loc='best')
plt.show()

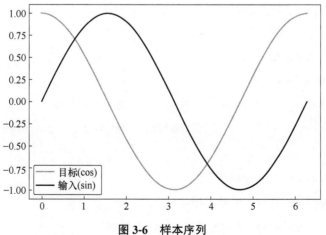

图 3-6　样本序列

秘籍 3-10 使用 PyTorch 内置的循环神经网络函数

问题

如何使用 PyTorch 设置循环神经网络模块并调用循环神经网络的函数?

解决方案

借助神经网络模块提供的内置函数,便可构建循环神经网络模型。

编程实战

以下示例展示了在 PyTorch 库中的神经网络模块中如何使用循环神经网络函数。在此脚本中,将设定输入矩阵的尺寸、隐藏层中的神经元数量以及网络中的隐藏层层数。

In [99]:
```
class RNN(nn.Module):
    def __init__(self):
        super(RNN, self).__init__()
        self.rnn = nn.RNN(
            input_size=INPUT_SIZE,
            hidden_size=32,      # 循环神经网络隐藏单元个数
            num_layers=1,        # 循环神经网络层数
            batch_first=True,    # 输入和输出将包含批处理大小维度为1的数据如
                                 # (batch, time_step, input_size)
        )
        self.out = nn.Linear(32, 1)
    def forward(self, x, h_state):
        # x 尺寸为 (batch, time_step, input_size)
```

```
        #h_state 尺寸为(n_layers, batch, hidden_size)
        #r_out 尺寸为(batch, time_step, hidden_size)
        r_out, h_state = self.rnn(x, h_state)
        outs = []                              # 保存所有预测结果
        for time_step in range(r_out.size(1)):    # 在每个时间步长计算输出
            outs.append(self.out(r_out[:, time_step, :]))
        return torch.stack(outs, dim=1), h_state
```

在定义循环神经网络类函数之后，还需提供优化算法，本例中选用 Adam 优化器。同时，损失函数采用均方损失函数。由于目标是对连续变量进行预测，因此在优化层中使用 MSELoss 函数作为损失函数。

In [100]:
```
rnn = RNN()
print(rnn)
RNN(
  (rnn): RNN(1, 32, batch_first=True)
  (out): Linear(in_features=32, out_features=1, bias=True)
)
```

In [101]:
```
optimizer = torch.optim.Adam(rnn.parameters(), lr=LR)   # 优化所有循环神经
                                                        网络参数

loss_func = nn.MSELoss()
```

In [102]:
```
h_state = None        # 初始隐藏状态
```

In [103]:
```
plt.figure(1, figsize=(12, 5))
plt.ion()              # 连续绘制图像
```

In [104]:
```
for step in range(60):
    start, end = step * np.pi, (step+1)*np.pi      # 时间范围
    # 使用 sin 函数预测 cos 函数值
    steps = np.linspace(start, end, TIME_STEP, dtype=np.float32)
    x_np = np.sin(steps)          # float32 转换为 torch FloatTensor
```

```
y_np = np.cos(steps)
x = Variable(torch.from_numpy(x_np[np.newaxis, :, np.newaxis]))
# 数据形状为（batch, time_step, input_size）
y = Variable(torch.from_numpy(y_np[np.newaxis, :, np.newaxis]))
prediction, h_state = rnn(x, h_state)      # rnn 的输出结果
# !! 下一步至关重要 !!
h_state = Variable(h_state.data)
# 重新打包隐藏状态，断开与上一次迭代的连接
loss = loss_func(prediction, y)            # 交叉熵损失
optimizer.zero_grad()                      # 清除此训练步骤的梯度
loss.backward()                            # 反向传播，计算梯度
optimizer.step()                           # 通过梯度下降执行一步参数更新
# 绘制结果
plt.plot(steps, y_np.flatten(), 'r-')
plt.plot(steps, prediction.data.numpy().flatten(), 'b-')
plt.draw(); plt.pause(0.05)
```

接下来，需要迭代 60 步，通过使用 sin 函数对从样本空间生成的 cos 函数进行预测。在迭代过程中，采用之前设定的学习率，并通过反向传播误差来降低均方误差（MSE），从而优化预测结果。迭代过程如 3-7 所示。其中黑色为 cos 函数的真值，灰色为预测值。

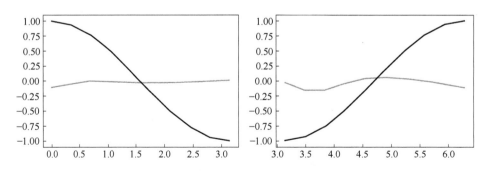

图 3-7　循环神经网络迭代优化的过程可视化

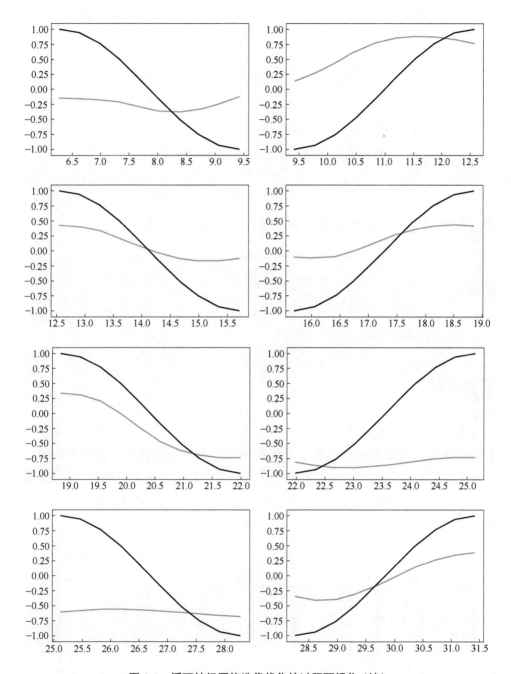

图 3-7 循环神经网络迭代优化的过程可视化（续）

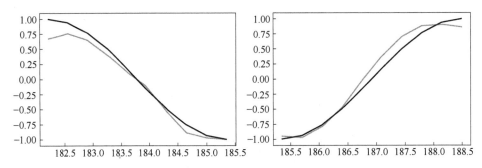

图 3-7　循环神经网络迭代优化的过程可视化（续）

秘籍 3-11　使用自编码器（Autoencoder）

问题

如何使用**自编码器**函数进行聚类？

解决方案

无监督学习作为机器学习的一个分支，其主要特点在于**不需要目标列**或**无输出定义**。我们的任务仅限于理解数据中存在的独特模式。如图 3-8 所示，在自编码器（Autoencoder）架构中，输入特征空间经过隐藏层转换至较低维度的张量表示，并映射回相同的输入空间。恰好位于中间的那一层保存了自编码器的值。

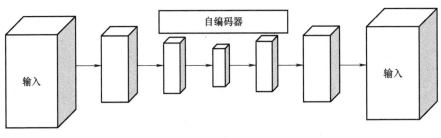

图 3-8　自编码器结构

编程实战

以下是一个实例。在 torchvision 库中,汇集了众多热门的数据集、模型架构以及实现框架。**自编码器**是一种从数据集中发掘潜在特征的方法,一般应用于分类、预测和聚类。在操作过程中,我们将输入数据置于输入层,将相同的数据集置于输出层,并添加多个隐藏层,每个隐藏层均包含众多神经元。经过多次迭代训练,最终在最内部的隐藏层中获取到一组潜在特征。中间隐藏层中的权重或参数被称为**自编码器层**。

In [105]:
```
# 自编码器
```

In [106]:
```
import torch
import torch.nn as nn
from torch.autograd import Variable
import torch.utils.data as Data
import torchvision
import matplotlib.pyplot as plt
from mpl_toolkits.mplot3d import Axes3D
from matplotlib import cm
import numpy as np
%matplotlib inline
```

In [107]:
```
torch.manual_seed(1)          # 为了可复现
```

In [108]:
```
# 超参数设置
EPOCH = 10
BATCH_SIZE = 64
LR = 0.005                    # 学习率
DOWNLOAD_MNIST = False
N_TEST_IMG = 5
```

在本实例中,我们依旧选用 MNIST 数据集以评估自编码器的性能。设置 10 个训练周期,每个批次包含 64 个输入样本,学习率为 0.005。同时,选取 5 张图像进行测试。

In [109]:

```
# Mnist手写数字数据集
train_data = torchvision.datasets.MNIST(
    root='./mnist/',
    train=True,
    # 表示这是训练数据
    transform=torchvision.transforms.ToTensor(),
    # 将PIL.Image类型或numpy.ndarray类型转换为形状为(C x H x W)的
      torch.FloatTensor类型，并在[0.0, 1.0]范围内进行归一化
    download=DOWNLOAD_MNIST,
    # 如果没有数据集，则需要进行下载
)
```

图 3-9 展示了从 torchvision 库加载的 MNIST 数据集，并以图像形式展示。

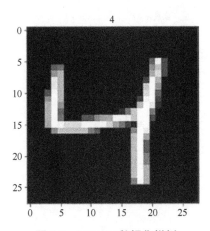

图 3-9 MNIST 数据集样例

In [110]:

```
# 绘制一个数据样例
print(train_data.train_data.size())         # 输入数据尺寸为(60000, 28, 28)
print(train_data.train_labels.size())       # 标签数据尺寸为(60000)
plt.imshow(train_data.train_data[2].numpy(), cmap='gray')
plt.title('%i' % train_data.train_labels[2])
```

```
plt.show()
torch.Size([60000, 28, 28])
torch.Size([60000])
```

In [111]:
```
# 数据加载器用于在训练中返回小批量数据，图像批次的形状将为
#(50, 1, 28, 28)
train_loader = Data.DataLoader(dataset=train_data, batch_size=BATCH_SIZE, shuffle=True)
```

In [112]:
```
class AutoEncoder(nn.Module):
    def __init__(self):
        super(AutoEncoder, self).__init__()
        self.encoder = nn.Sequential(
            nn.Linear(28*28, 128),
            nn.Tanh(),
            nn.Linear(128, 64),
            nn.Tanh(),
            nn.Linear(64, 12),
            nn.Tanh(),
            nn.Linear(12, 3),   # 压缩为可在 plt 中可视化的 3 个特征
        )
        self.decoder = nn.Sequential(
            nn.Linear(3, 12),
            nn.Tanh(),
            nn.Linear(12, 64),
            nn.Tanh(),
            nn.Linear(64, 128),
            nn.Tanh(),
            nn.Linear(128, 28*28),
            nn.Sigmoid(),                # 压缩到(0,1)范围
        )

    def forward(self, x):
        encoded = self.encoder(x)
        decoded = self.decoder(encoded)
        return encoded, decoded
```

下面，我们来探讨自编码器的结构：
- 输入层：包含 784 个特征，对应于 28x28 像素的图像。
- 第一个隐藏层：包含 128 个神经元，接收来自输入层的 784 个特征，并应用双曲正切函数传递信息到下一层。
- 第二个隐藏层：同样包含 128 个神经元，将信息转换为 64 个特征。
- 第三个隐藏层：再次应用双曲正切函数传递信息到下一层。
- 最内层（编码器的最后一层）：包含 3 个神经元，这些神经元被视为 3 个特征，即编码的潜在表示。
- 解码器：将内部层的特征扩展回 784 个特征，以重建原始输入图像。

In [113]:

```
autoencoder = AutoEncoder()
print(autoencoder)

optimizer = torch.optim.Adam(autoencoder.parameters(), lr=LR)
loss_func = nn.MSELoss()
# 查看原始数据(第一行)
view_data = Variable(train_data.train_data[:N_TEST_IMG].view(-1, 28*28).type(torch.FloatTensor)/255.)

AutoEncoder(
  (encoder): Sequential(
    (0): Linear(in_features=784, out_features=128, bias=True)
    (1): Tanh()
    (2): Linear(in_features=128, out_features=64, bias=True)
    (3): Tanh()
    (4): Linear(in_features=64, out_features=12, bias=True)
    (5): Tanh()
    (6): Linear(in_features=12, out_features=3, bias=True)
  )
  (decoder): Sequential(
    (0): Linear(in_features=3, out_features=12, bias=True)
    (1): Tanh()
    (2): Linear(in_features=12, out_features=64, bias=True)
    (3): Tanh()
    (4): Linear(in_features=64, out_features=128, bias=True)
    (5): Tanh()
```

```
    (6): Linear(in_features=128, out_features=784, bias=True)
    (7): Sigmoid()
  )
)
```

在确定模型架构后，需通过设定学习率和优化函数，以最小化损失函数。整个架构将经过多个批次的训练迭代，直至达到目标输出。

秘籍 3-12　使用自编码器实现结果微调

问题

如何设置迭代次数来进行微调结果？

解决方案

在理论层面，自编码器与聚类模型的运作方式具有相似性。在无监督学习过程中，机器能够从数据中挖掘潜在模式，并将其应用于新的数据集。此过程主要通过捕获一系列输入特征来实现。自编码器函数在很大程度上也应用于特征工程领域。

编程实战

我们以一个实例来进行说明。在此，仍然选择 MNIST 数据集，目的是深入探讨在优化自编码器层时 epoch 值的作用。可以通过增加 epoch 值来最小化误差；然而，在实际操作中，增加 epoch 值会面临诸如内存限制等挑战。训练过程的可视化结果如图 3-10 所示。

In [114]:
```
for epoch in range(EPOCH):
    for step, (x, y) in enumerate(train_loader):
        b_x = Variable(x.view(-1, 28*28))   # 批数据 b_x, 形状为（batch, 28*28）
        b_y = Variable(x.view(-1, 28*28))   # 批数据 b_y, 形状为（batch, 28*28）
```

```python
        b_label = Variable(y)                    # 批标签
        encoded, decoded = autoencoder(b_x)
        loss = loss_func(decoded, b_y)           # 均方误差作为损失函数
        optimizer.zero_grad()                    # 清除此训练步骤的梯度
        loss.backward()                          # 反向传播,计算梯度
        optimizer.step()                         # 通过梯度下降执行一步参数更新
        if step % 500 == 0 and epoch in [0, 5, EPOCH-1]:
            print('Epoch: ', epoch, '| train loss: %.4f' % loss.data)
            # 绘制解码后的图像(第二行)
            _, decoded_data = autoencoder(view_data)
            # 初始化图形
            f, a = plt.subplots(2, N_TEST_IMG, figsize=(5, 2))
            for i in range(N_TEST_IMG):
                a[0][i].imshow(np.reshape(view_data.data.numpy()[i],
                                (28, 28)), cmap='gray');
                a[0][i].set_xticks(()); a[0][i].set_yticks(())
            for i in range(N_TEST_IMG):
                a[1][i].clear()
                a[1][i].imshow(np.reshape(decoded_data.data.numpy()[i],
                                (28, 28)), cmap='gray')
                a[1][i].set_xticks(()); a[1][i].set_yticks(())
            plt.show(); #plt.pause(0.05)
```

图 3-10　训练过程可视化

轮次：5 | 训练误差：0.0375

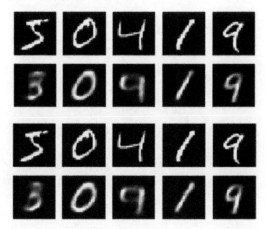

轮次：9 | 训练误差：0.0378

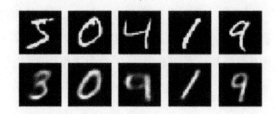

轮次：9 | 训练误差：0.0382

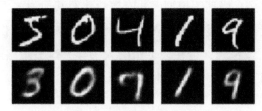

图 3-10　训练过程可视化（续）

利用编码器函数，将输入特征转化为潜在特征。同时，通过应用解码器函数，可以实现图像的重构。进一步地，借助自编码器函数，可以匹配图像重构的方式。从图中序列可以看出，随着训练轮次的增加，解码重构的图像明显变得更加易于识别。

秘籍 3-13　约束模型过拟合

问题

在运用大量神经元和层进行目标类别或输出变量的预测时，模型往往容易对训练数据集产生过拟合现象。这种过拟合导致我们在测试集上的预测表现不佳。因此，测试准确率与训练准确率之间存在差异，且两者之间偏差较大。

解决方案

为了约束模型过拟合，引入丢弃率（Dropout rate）的策略，即在网络中随机裁剪如 10% 或 20% 的权重，并同时检测模型的准确性。若在权重裁剪至 10% 或 20% 后，模型仍能保持相同的准确性，则说明该模型表现良好。

编程实战

以下是一个实例。当训练模型无法有效泛化至其他测试用例场景时，便会产生过拟合现象。过拟合的显著特征表现为训练准确率与测试准确率之间存在较大偏差。为防止模型过拟合，可在模型中引入丢弃率这一策略。图 3-11 所示为模型训练过程中的训练准确率和测试准确率。

In [115]:
```
import torch
from torch.autograd import Variable
import matplotlib.pyplot as plt
%matplotlib inline

torch.manual_seed(1)   # 为了可复现
```

In [116]:
```
N_SAMPLES = 20
N_HIDDEN = 300
```

In [117]:
训练数据
```
x = torch.unsqueeze(torch.linspace(-1, 1, N_SAMPLES), 1)
y = x + 0.3*torch.normal(torch.zeros(N_SAMPLES, 1), torch.ones(N_SAMPLES, 1))
x, y = Variable(x), Variable(y)
```

In [118]:
测试数据
```
test_x = torch.unsqueeze(torch.linspace(-1, 1, N_SAMPLES), 1)
test_y = test_x + 0.3*torch.normal(torch.zeros(N_SAMPLES, 1), torch.ones(N_SAMPLES, 1))
test_x, test_y = Variable(test_x), Variable(test_y )
```

In [119]:
显示数据
```
plt.scatter(x.data.numpy(), y.data.numpy(), c='magenta', s=50, alpha=0.5, label='train')
plt.scatter(test_x.data.numpy(), test_y.data.numpy(), c='cyan', s=50, alpha=0.5, label='test')
plt.legend(loc='upper left')
plt.ylim((-2.5, 2.5))
plt.show()
```

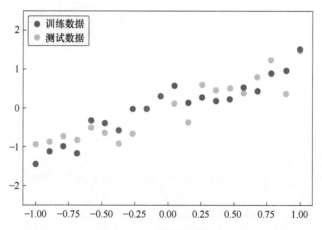

图 3-11　模型训练过程中的训练准确率和测试准确率

丢弃率机制的引入确保了当隐藏层中的权重低于预设阈值时，相应的神经元将被移除。在实际应用中,丢弃率通常介于 20%~50% 之间。20% 的丢弃率意味着对模型权重的惩罚相对较小;然而,50% 的阈值则表示对模型权重的严格限制。

在以下脚本中,我们对模型权重实施了 50% 的丢弃率,以实现权重的剔除。在该模型中,共应用了两次丢弃率处理。

In [120]:

```
net_overfitting = torch.nn.Sequential(
    torch.nn.Linear(1, N_HIDDEN),
    torch.nn.ReLU(),
    torch.nn.Linear(N_HIDDEN, N_HIDDEN),
    torch.nn.ReLU(),
    torch.nn.Linear(N_HIDDEN, 1),
)
```

In [121]:

```
net_dropped = torch.nn.Sequential(
    torch.nn.Linear(1, N_HIDDEN),
    torch.nn.Dropout(0.5),   # 丢弃 50% 的神经元
    torch.nn.ReLU(),
    torch.nn.Linear(N_HIDDEN, N_HIDDEN),
    torch.nn.Dropout(0.5),   # 丢弃 50% 的神经元
    torch.nn.ReLU(),
    torch.nn.Linear(N_HIDDEN, 1),
)
```

In [122]:

```
print(net_overfitting)      # 网络架构
print(net_dropped)

Sequential(
  (0): Linear(in_features=1, out_features=300, bias=True)
  (1): ReLU()
  (2): Linear(in_features=300, out_features=300, bias=True)
  (3): ReLU()
  (4): Linear(in_features=300, out_features=1, bias=True)
)
Sequential(
  (0): Linear(in_features=1, out_features=300, bias=True)
```

```
    (1): Dropout(p=0.5, inplace=False)
    (2): ReLU()
    (3): Linear(in_features=300, out_features=300, bias=True)
    (4): Dropout(p=0.5, inplace=False)
    (5): ReLU()
    (6): Linear(in_features=300, out_features=1, bias=True)
)
```

In [123]:
```
optimizer_ofit = torch.optim.Adam(net_overfitting.parameters(), lr=0.01)
optimizer_drop = torch.optim.Adam(net_dropped.parameters(), lr=0.01)
loss_func = torch.nn.MSELoss()
```

在上述代码中，我们构建了两个神经网络：net_overfitting 与 net_dropped。其中，net_dropped 网络采用了 Dropout 层，该层会随机剔除 50% 的神经元。这种做法有助于防止过拟合现象，增强模型的泛化能力。同时，我们还初始化了优化器和损失函数，为训练模型做好准备。

需要注意的是，选择合适的丢弃率需要对业务和领域有一定的了解。

秘籍 3-14　可视化模型过拟合

问题

评估深度学习模型的过拟合。

解决方案

调整模型中的超参数，随后持续优化，观察模型是否出现对数据过拟合的现象。

编程实战

在上一节中，我们构建了两种类型的神经网络：过拟合网络（net_overfitting）

和减连接网络（net_dropped）。当模型参数从训练数据中拟合出来，且越来越接近训练数据集的实际数据，而同样的模型在测试集上表现有所不同，这就是模型过拟合的明显迹象。为了抑制模型过拟合现象，可以引入丢弃率技术，它通过删除一定比例的连接（即网络中的权重）来使训练好的模型更接近真实数据。

在下列脚本中，我们将迭代次数设定为500次。分别展示基础模型（net_overfitting）产生的过拟合预测结果以及Dropout模型（net_dropped）生成的预测数据。采用同样的方法，分别创建了两套损失函数、反向传播算法以及优化器。如图3-12所示为两个模型的拟合结果。

In [124]:
```python
for t in range(500):
    pred_ofit = net_overfitting(x)
    pred_drop = net_dropped(x)
    loss_ofit = loss_func(pred_ofit, y)
    loss_drop = loss_func(pred_drop, y)

    optimizer_ofit.zero_grad()
    optimizer_drop.zero_grad()
    loss_ofit.backward()
    loss_drop.backward()
    optimizer_ofit.step()
    optimizer_drop.step()

    if t % 100 == 0:
    # 切换到评估模式以修复dropout效果
    net_overfitting.eval()

    net_dropped.eval()   #dropout的参数与训练模式不同
    # 绘图
    plt.cla()
    test_pred_ofit = net_overfitting(test_x)
    test_pred_drop = net_dropped(test_x)
    plt.scatter(x.data.numpy(), y.data.numpy(), c='magenta', s=50,
                alpha=0.3, label='train')
    plt.scatter(test_x.data.numpy(), test_y.data.numpy(),
    c='cyan', s=50,
                alpha=0.3, label='test')
    plt.plot(test_x.data.numpy(), test_pred_ofit.data.numpy(), 'r-',
```

```
          lw=3, label='overfitting')
plt.plot(test_x.data.numpy(), test_pred_drop.data.numpy(), 'b--',
          lw=3, label='dropout(50%)')
plt.text(0, -1.2, 'overfitting loss=%.4f' % loss_func(test_pred_
ofit, test_y).data,
          fontdict={'size': 20, 'color':  'red'})
plt.text(0, -1.5, 'dropout loss=%.4f' % loss_func(test_pred_drop,
test_y).data,
          fontdict={'size': 20, 'color': 'blue'})
plt.legend(loc='upper left'); plt.ylim((-2.5, 2.5));plt.pause(0.1)
# 恢复到训练模式
net_overfitting.train()
net_dropped.train()
plt.show()
```

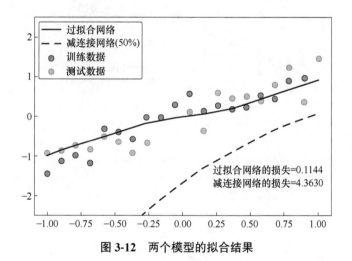

图 3-12　两个模型的拟合结果

如图 3-13 所示，图像中绘制的内容包括过拟合网络的损失和减连接网络的损失，以及它们与实际训练和测试数据点的差异。

经过多轮优化，我们采用 net_overfitting 与 net_dropped 两个模型函数生成了上述图表。从图表分析来看，**训练数据**可能更贴近过拟合模型（net_overfitting）；然而，net_dropped 模型在拟合整体数据方面表现更为优异。

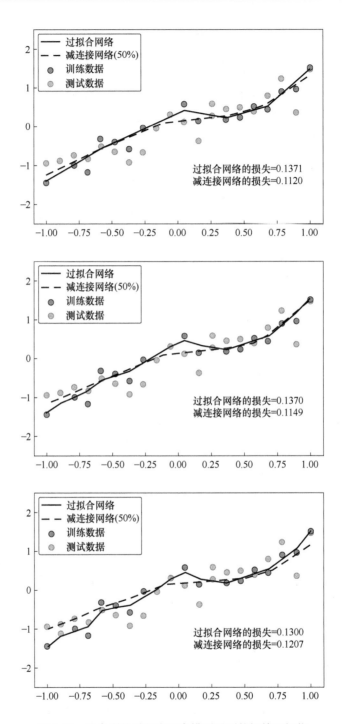

图 3-13 拟合训练过程中两个模型预测数据的可视化

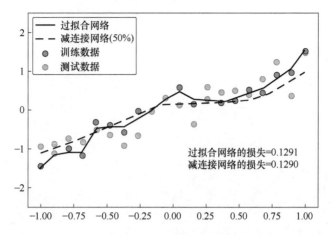

图 3-13　拟合训练过程中两个模型预测数据的可视化（续）

秘籍 3-15　初始化权重中的丢弃率

问题

在网络中如何剔除权重？应该是随机剔除还是使用分布进行剔除？

解决方案

在剔除 Dropout 层中的权重时，应依据概率分布进行，而非随机剔除。

编程实战

在先前的示例中，我们所使用的模型包含了三个 Dropout 层：第一个位于第一个隐藏层之后，第二个和第三个位于第二个隐藏层之后。丢弃率的概率百分比为 0.50，即意味着有 50% 的权重会被随机剔除。然而，在某些情况下，随机选择并删除网络中的权重可能会影响到一些相关的权重。为此，我们可以考虑采用另一种可行的替代方法——根据统计分布来剔除网络中的权重。

以下脚本展示了如何从均匀分布中生成权重，进而将这些权重应用于网络架构之中。

In [125]:
初始化一个随机张量

In [126]:
import numpy as np

In [127]:
import torch
# 来自均匀分布

In [128]:
torch.Tensor(5, 3)

Out[128]:
tensor([[2.6019e-33, 0.0000e+00, 3.7835e-44], [0.0000e+00, nan, 0.0000e+00], [1.3733e-14, 6.4069e+02, 4.3066e+21], [1.1824e+22, 4.3066e+21, 6.3828e+28], [3.8016e-39, 0.0000e+00, 1.5501e-37]])

In [129]:
使用均匀分布初始化随机张量

In [130]:
torch.Tensor(5, 3).uniform_(-1, 1)

Out[130]:
tensor([[0.8790, 0.7375, 0.1182], [0.3652, 0.1322, 0.8587], [0.3682, -0.2907, 0.0051], [0.0886, -0.7588, -0.5371], [0.0085, 0.0812, -0.6360]])

In [131]:
获取张量的形状

In [132]:
x = torch.Tensor(5, 3).uniform_(-1, 1)
print(x.size())
torch.Size([5, 3])

In [133]:
使用 list 和 numpy 类型的数据创建张量

In [134]:
z = torch.LongTensor([[1, 3], [2, 9]])
print(z.type())

```
# 转换为numpy ndarray
print(z.numpy().dtype)
torch.LongTensor
int64
In [135]:
# 从numpy类型创建的数据,判断其数据类型
print(torch.from_numpy(np.random.rand(5, 3)).type())
print(torch.from_numpy(np.random.rand(5, 3).astype(np.float32)).type())
torch.DoubleTensor
torch.FloatTensor
```

秘籍3-16　添加数学运算

问题

如何设置广播函数并优化卷积函数？

解决方案

本节中的代码将演示如何在构建卷积神经网络模型时引入批量归一化（Batch Normalization）技术，并在其后设置池化层。

编程实战

在引入神经网络模型的卷积层中进行批量归一化过程中，需执行基于张量的数学运算，这些运算在功能上与其他计算方法有所区别。

```
In [136]:
# 简单的数学运算
In [137]:
y = x * torch.randn(5, 3)
print(y)
```

```
tensor([[ 0.1587,  0.4137, -0.4801],
        [-0.2706,  0.0411, -0.8954],
        [ 0.3616, -0.0245, -0.3401],
        [-0.6478, -0.1207, -0.1698],
        [ 0.2107, -0.2128,  0.1017]])
```

In [138]:

```
y = x / torch.sqrt(torch.randn(5, 3) ** 2)
print(y)
tensor([[ 2.1697, -1.1561, -7.4875],
        [-0.5094,  0.4193, -4.4016],
        [ 0.4308,  0.0421,  0.6234],
        [ 2.3634,  2.1020, -0.2185],
        [ 4.8023,  0.4352,  0.4892]])
```

In [139]:

```
# 广播机制
```

In [140]:

```
print (x.size())
y = x + torch.randn(5, 1)
print(y)
torch.Size([5, 3])
tensor([[ 1.0416, -0.1192, -0.1256],
        [ 0.0484,  1.5687,  0.0468],
        [ 0.1000, -0.4971,  0.3657],
        [ 0.3893,  0.5367, -0.2656],
        [ 2.1538,  1.9121,  2.0349]])
```

In [141]:

```
# 重塑形状
```

In [142]:

```
y = torch.randn(5, 10, 15)
print(y.size())
print(y.view(-1, 15).size())    # 等同于执行 y.view(50, 15)
print(y.view(-1, 15).unsqueeze(1).size())   # 在索引 1 处添加一个维度
print(y.view(-1, 15).unsqueeze(1).squeeze().size())
print()
print(y.transpose(0, 1).size())
print(y.transpose(1, 2).size())
```

```
print(y.transpose(0, 1).transpose(1, 2).size())
print(y.permute(1, 2, 0).size())
torch.Size([5, 10, 15])
torch.Size([50, 15])
torch.Size([50, 1, 15])
torch.Size([50, 15])

torch.Size([10, 5, 15])
torch.Size([5, 15, 10])
torch.Size([10, 15, 5])
torch.Size([10, 15, 5])
```

In [143]:

重复

In [144]:

```
print(y.view(-1, 15).unsqueeze(1).expand(50, 100, 15).size())
print(y.view(-1, 15).unsqueeze(1).expand_as(torch.randn(50, 100, 15)).size())
torch.Size([50, 100, 15])
torch.Size([50, 100, 15])
```

In [145]:

拼接张量

In [146]:

```
# 2 是张量连接的维度
print(torch.cat([y, y], 2).size())
# stack 沿着新维度将张量序列连接起来
print(torch.stack([y, y], 0).size())

torch.Size([5, 10, 30])
torch.Size([2, 5, 10, 15])
```

In [147]:

高级索引

In [148]:

```
y = torch.randn(2, 3, 4)
print(y[[1, 0, 1, 1]].size())
# PyTorch 目前不支持负步长，因此 -1 不起作用
```

```
rev_idx = torch.arange(1, -1, -1).long()
print(y[rev_idx].size())
```

```
torch.Size([4, 3, 4])
torch.Size([2, 3, 4])
```

In [149]:
卷积层、批量归一化层和池化层

In [150]:
```
x = Variable(torch.randn(10, 3, 28, 28))

conv = nn.Conv2d(in_channels=3, out_channels=32, kernel_size=(3, 3),
stride=1,
                 padding=1, bias=True)
bn = nn.BatchNorm2d(num_features=32)
pool = nn.MaxPool2d(kernel_size=(2, 2), stride=2)

output_conv = bn(conv(x))
outpout_pool = pool(conv(x))

print('Conv output size : ', output_conv.size())
print('Pool output size : ', outpout_pool.size())
Conv output size :   torch.Size([10, 32, 28, 28])
Pool output size :   torch.Size([10, 32, 14, 14])
```

秘籍 3-17　循环神经网络中的嵌入层

问题

　　循环神经网络在文本处理领域具有广泛应用，其相较于原始特征，利用嵌入特征能够提升标准循环神经网络模型的准确性。然而，如何在循环神经网络中构建嵌入特征以提高模型性能成了一个关键问题。

解决方案

　　首先，需要构建一个嵌入层（Embedding Layer），它是由一个固定字典和固定大小的查找表组成的。接着，在创建门控循环单元（Gated Recurrent Unit，GRU）之前，需要引入丢弃率。

编程实战

在文本数据以序列形式输入时，信息会按顺序进行处理。例如，在描述某一事物时，我们通常会通过一系列有序的词汇来传达含义。若采用单一单词作为数据表示的向量，所得到的数据集将呈现高度稀疏性。然而，通过运用基于短语的方法或将单词组合形成特征向量，这些向量将转变为密集的一层。这种密集向量层被称为词嵌入，其作用在于传达上下文或意义。相较于词袋模型，这种方法具有显著的优势。

In [151]:

```
# 循环层、嵌入层和 Dropout 层
```

In [152]:

```
inputs = [[1, 2, 3], [1, 0, 4], [1, 2, 4], [1, 4, 0], [1, 3, 3]]
x = Variable(torch.LongTensor(inputs))

embedding = nn.Embedding(num_embeddings=5, embedding_dim=20, padding_idx=1)
drop = nn.Dropout(p=0.5)
gru = nn.GRU(input_size=20, hidden_size=50, num_layers=2, batch_first=True,
            bidirectional=True, dropout=0.3)

emb = drop(embedding(x))
gru_h, gru_h_t = gru(emb)

print('Embedding size : ', emb.size())
print('GRU hidden states size : ', gru_h.size())
print('GRU last hidden state size : ', gru_h_t.size())
Embedding size :   torch.Size([5, 3, 20])
GRU hidden states size :   torch.Size([5, 3, 100])
GRU last hidden state size :   torch.Size([4, 5, 50])
```

In [153]:

```
# functional API 为用户提供以功能方式使用神经网络类的方法，即所调用的函数没有
    可学习的参数
```

In [154]:
```
import torch.nn.functional as F
x = Variable(torch.randn(10, 3, 28, 28))
filters = Variable(torch.randn(32, 3, 3, 3))
conv_out = F.relu(F.dropout(F.conv2d(input=x, weight=filters, padding=1),
                            p=0.5, training=True))

print('Conv output size : ', conv_out.size())
Conv output size :   torch.Size([10, 32, 28, 28])
```

小结

本章介绍了如何运用 PyTorch 的 API 构建简单的神经网络模型，并通过调整超参数（如学习率、迭代次数及梯度衰减等）以优化模型参数。此外，我们还掌握了创建卷积神经网络和循环神经网络的方法，并在这些网络结构中引入 Dropout 以抑制过拟合现象。

在本章中，我们采用小型张量来追踪计算及其他过程的细节。只需明确问题陈述，构建特征，随后应用各节中的代码，即可获得相应的结果。下一章将展示使用 PyTorch 实现的更多实例。

第 4 章

PyTorch 中的神经网络简介

深度神经网络模型已逐渐成为人工智能与机器学习领域的核心实现技术。基于人工神经网络的高级建模技术正引领数据挖掘走向未来。然而，一个不容忽视的问题是，神经网络自 20 世纪 50 年代问世以来，为何直至如今才获得广泛关注与重视呢？

神经网络作为一种并行信息处理系统，起源于计算机科学领域。其架构模拟了人类大脑中神经元的相互关联特性，用于信息的传递与处理，从而实现人脸识别、图像识别等功能。在本章中，我们将探讨如何将神经网络方法应用于各类数据挖掘任务，如分类、回归、预测以及特征降维等。人工神经网络（Artificial Neural Network，ANN）模拟了人类大脑的运作方式，通过数十亿个神经元的相互连接，实现信息处理和洞察生成。

秘籍 4-1 激活函数的使用

问题

激活函数是什么，它们在实际项目中如何工作？如何使用 PyTorch 实现激活函数？

解决方案

激活函数是一种数学公式,根据所应用的数学变换函数类型,可以对二进制、浮点或整数类型的向量进行变换。在神经网络中,神经元分布于不同的层级,包括输入层、隐藏层和输出层,这些层级通过激活函数这种数学函数相互连接。激活函数有多种形式,以下将逐一阐述。深入理解激活函数对于精确实现神经网络模型至关重要。

编程实战

在神经网络模型中,激活函数可分为线性函数与非线性函数两大类。在 PyTorch 中使用 torch.nn 模块可以创建各种类型的神经网络模型。以下将展示利用 PyTorch 及 torch.nn 模块实现激活函数部署的若干实例。

PyTorch 与 TensorFlow 之间的核心差异主要体现在计算图的定义方式、执行计算的方法以及在不同脚本调整和引入其他基于 Python 库的灵活性。在 TensorFlow 中,用户需要在初始化模型之前明确定义变量和占位符,同时需关注后续所需的对象,因此需要采用一个占位符(placeholder)。另外在 TensorFlow 中,用户需先定义模型,然后再进行编译和运行;相较之下,PyTorch 允许用户随时定义模型,无需在代码中保留占位符,这正是 PyTorch 框架被称为动态的原因。

线性函数

线性函数作为一种简洁的数学关系,常被应用于将信息由解映射层(demapping layer)传输至输出层。在处理变化较为平缓的数据时,线性函数往往成为首选。在深度学习模型中,从最后一个隐藏层到输出层,通常也使用线性函数。线性函数的特点在于,输出值始终保持在特定范围内,因此在深度学习模型的最后一个隐藏层中、基于线性回归的任务中以及从输入数据预测结果的深度学习模型中,均可以见到线性函数的身影。其公式为

$$y = a + \beta x$$

双线性函数

双线性函数是一种简单的函数,通常用于传递信息。它对输入数据应用双线性变换。其公式为

$$y = x_1 * A * x_2 + b$$

In [1]:
```
from __future__ import print_function
import torch
import numpy as np
import torch.optim
import torch.nn as nn
import torch.optim as optim
import torch.nn.init as init
import torch.nn.functional as F
from torch.autograd import Variable
```

In [2]:
```
import warnings
warnings.filterwarnings("ignore", category=FutureWarning)
```

In [3]:
```
# 可以使用torch.nn包来构建神经网络
```

In [4]:
```
x = Variable(torch.randn(100, 10))
y = Variable(torch.randn(100, 30))

linear = nn.Linear(in_features=10, out_features=5, bias=True)
output_linear = linear(x)
print('Output size : ', output_linear.size())

bilinear = nn.Bilinear(in1_features=10, in2_features=30, out_features=5, bias=True)
output_bilinear = bilinear(x, y)
print('Output size : ', output_bilinear.size())
Output size :  torch.Size([100, 5])
Output size :  torch.Size([100, 5])
```

Sigmoid 函数

鉴于其易于解释和实现的特性，Sigmoid 函数在数据挖掘和分析领域得到了广泛专业人士的运用。作为一种非线性函数，它在神经网络中具有重要作用。当权重从输入层传输至隐藏层时，我们希望模型能够捕捉到数据中所蕴含的各种非线性关系。因此，在神经网络的隐藏层中运用 Sigmoid 函数是推荐的做法。非线性函数有助于实现数据集的泛化，并使得函数梯度的计算更为简便。

Sigmoid 函数是一种特定的非线性激活函数，其输出值始终限定在 0~1 之间。因此，该函数广泛应用于基于分类任务的处理。然而，Sigmoid 函数的一个潜在缺陷是可能陷入局部最小值。但是，它的优点在于提供了属于某一类别的概率。Sigmoid 函数的公式为

$$f(x) = \frac{1}{1+e^{-\beta x}}$$

In [5]:
```
x = Variable(torch.randn(100, 10))
y = Variable(torch.randn(100, 30))

sig = nn.Sigmoid()
output_sig = sig(x)
output_sigy = sig(y)
print('Output size : ', output_sig.size())
print('Output size : ', output_sigy.size())
```
```
Output size :  torch.Size([100, 10])
Output size :  torch.Size([100, 30])
```
In [6]:
```
print(x[0])
print(output_sig[0])
```
```
tensor([-1.5454,  0.3599,  2.2720,  0.7115,  0.5296,  0.6176,  1.8854,  0.4854,
        -0.3893,  0.8369])
tensor([0.1758, 0.5890, 0.9065, 0.6707, 0.6294, 0.6497, 0.8682, 0.6190,
0.4039, 0.6978])
```

双曲正切函数

双曲正切函数作为一种变换函数的变体，在神经网络模型中，常用于将从映射层获取的信息传递至隐藏层。该函数在隐藏层之间发挥着重要作用。其取值范围介于 –1~1 之间，公式为：

$$\tanh(x) = \frac{e^x - e^{-x}}{e^x + e^{-x}}$$

In[7]:
```
x = Variable(torch.randn(100, 10))
y = Variable(torch.randn(100, 30))

func = nn.Tanh()
output_x = func(x)
output_y = func(y)
print('Output size : ', output_x.size())
print('Output size : ', output_y.size())

Output size :  torch.Size([100, 10])
Output size :  torch.Size([100, 30])
```
In[8]:
```
print(x[0])
print(output_x[0])
print(y[0])
print(output_y[0])

tensor([ 1.6056,  0.1092,  0.2044,  1.0537, -0.8658, -0.9111, -1.1586, -1.7745,
        -0.8922, -2.3219])
tensor([ 0.9225,  0.1087,  0.2016,  0.7832, -0.6992, -0.7217, -0.8206, -0.9441,
        -0.7125, -0.9809])
tensor([ 0.2153,  1.3900,  0.4259, -0.3347, -1.2087, -0.1930,  0.1645, -1.5867,
        -0.1752,  0.3863,  0.6141,  1.6769, -0.8080,  0.3790, -0.7446,  0.1795,
        -1.5132,  0.8282,  1.6872,  0.7207, -0.6874,  0.0136,  0.3600,  1.9525,
        -0.1363, -0.2002,  0.4026, -0.1413,  2.2343,  1.0469])
tensor([ 0.2121,  0.8832,  0.4019, -0.3228, -0.8363, -0.1907,  0.1631, -0.9196,
        -0.1735,  0.3682,  0.5470,  0.9325, -0.6685,  0.3619, -0.6319,  0.1776,
        -0.9075,  0.6795,  0.9338,  0.6173, -0.5963,  0.0136,  0.3452,  0.9605,
        -0.1355, -0.1976,  0.3821, -0.1404,  0.9773,  0.7806])
```

对数 Sigmoid 传递函数

对数 Sigmoid 传递函数适用于将输入层映射至隐藏层。如果数据不是二元的，而是浮点类型，且存在大量异常值（如输入特征中存在较大数值）时，则应该使用对数 Sigmoid 传递函数。该函数公式为

$$f(x) = \log\left(\frac{1}{1+e^{-\beta x}}\right)$$

In[9]:
```
x = Variable(torch.randn(100, 10))
y = Variable(torch.randn(100, 30))

func = nn.LogSigmoid()
output_x = func(x)
output_y = func(y)
print('Output size : ', output_x.size())
print('Output size : ', output_y.size())
```
```
Output size :  torch.Size([100, 10])
Output size :  torch.Size([100, 30])
```
In[10]:
```
print(x[0])
print(output_x[0])
print(y[0])
print(output_y[0])
```
```
tensor([-0.9983, -0.2337,  0.7794,  1.0399, -1.4705, -1.4177, -0.2531,
        -1.0391, -1.1570, -0.5105])
tensor([-1.3120, -0.8168, -0.3775, -0.3027, -1.6773, -1.6346, -0.8277, -1.3420,
        -1.4304, -0.9806])
tensor([-0.3758, -1.1889,  0.7846,  0.8277,  0.1351,  0.2677, -0.2810, -1.1610,
        -0.6973, -0.1106,  0.6361,  1.4497, -0.6007, -0.1102,  0.8876, -0.1440,
        -0.2914, -0.0144,  1.4152,  2.1429,  0.8828,  0.9561, -0.1876,  1.1487,
         0.6150, -0.1044,  1.3075, -0.1601, -0.4018, -1.2599])
tensor([-0.8986, -1.4547, -0.3759, -0.3626, -0.6279, -0.5683, -0.8435, -1.4335,
        -1.1014, -0.7500, -0.4249, -0.2108, -1.0379, -0.7498, -0.3448, -0.7677,
```

        -0.8494, -0.7004, -0.2174, -0.1109, -0.3462, -0.3253, -0.7913, -0.2754,
        -0.4322, -0.7467, -0.2394, -0.7764, -0.9141, -1.5096])

ReLU 函数

ReLU（Rectified Linear Unit，修正线性单元）函数也是一种激活函数，其主要作用在于将信息从输入层传递至输出层。在卷积神经网络模型中，ReLU 激活函数具有广泛应用。该激活函数的输出数值范围为 0~+∞。它通常在神经网络模型的不同隐藏层之间使用。

In [11]:
```
X = Variable(torch.randn(100, 10))
y = Variable(torch.randn(100, 30))

func = nn.ReLU()
output_x = func(x)
output_y = func(y)
print('Output size : ', output_x.size())
print('Output size : ', output_y.size())
```
Output size : torch.Size([100, 10])
Output size : torch.Size([100, 30])

In [12]:
```
print(x[0])
print(output_x[0])
print(y[0])
print(output_y[0])
```
tensor([-0.6479, -0.8856, 0.5144, -0.5064, 0.3280, -1.8378, 0.5670,
         0.9095, -2.6267, -1.0119])
tensor([0.0000, 0.0000, 0.5144, 0.0000, 0.3280, 0.0000, 0.5670, 0.9095,
        0.0000, 0.0000])
tensor([-1.4458, 0.8328, 0.6534, 2.0404, 0.9053, -0.2829, -0.5712, 0.0323,
         0.9757, -1.5787, 1.9665, 1.0276, -1.0536, 0.0588, 0.5085, 0.1956,
        -0.4490, -0.8927, 0.0128, -0.5971, -0.0677, 0.0101, 0.9477, 1.1218,
        -1.0648, -0.8439, 0.3422, 0.6930, -0.4311, -1.2920])
tensor([0.0000, 0.8328, 0.6534, 2.0404, 0.9053, 0.0000, 0.0000, 0.0323, 0.9757,
        0.0000, 1.9665, 1.0276, 0.0000, 0.0588, 0.5085, 0.1956, 0.0000, 0.0000,

0.0128, 0.0000, 0.0000, 0.0101, 0.9477, 1.1218, 0.0000, 0.0000, 0.3422, 0.6930, 0.0000, 0.0000])

在神经网络架构中，不同类型的传递函数具有可替换性。它们可应用于不同阶段，如由输入层至隐藏层，或由隐藏层至输出层，从而提升模型准确性。

Leaky ReLU

泄漏线性整流单元（Leaky ReLU）作为一种激活函数，旨在解决传统神经网络模型中普遍存在的梯度消失问题。梯度消失现象是指在反向传播过程中，梯度逐渐减小，使得权重更新几乎停滞，从而影响模型的训练效果。为解决此问题，引入了Leaky ReLU激活函数。Leaky ReLU在神经元不活跃时，允许存在一个较小且非零的梯度。

In [13]:

```
X = Variable(torch.randn(100, 10))
y = Variable(torch.randn(100, 30))

func = nn.LeakyReLU()
output_x = func(x)
output_y = func(y)
print('Output size : ', output_x.size())
print('Output size : ', output_y.size())

Output size :   torch.Size([100, 10])
Output size :   torch.Size([100, 30])
```

In [14]:

```
print(x[0])
print(output_x[0])
print(y[0])
print(output_y[0])
tensor([ 0.3611, -0.3622,  0.5740, -0.3404, -0.1284,  1.4639,  1.3272,
         0.0636, -1.1366,  1.1084])
tensor([ 3.6107e-01, -3.6216e-03,  5.7399e-01, -3.4043e-03, -1.2843e-03,
         1.4639e+00,  1.3272e+00,  6.3646e-02, -1.1366e-02,  1.1084e+00])
tensor([-0.4000, -0.2603,  0.5494, -1.1904,  1.0810,  0.0770,  0.5700, -1.0860,
         0.6954, -0.3596, -0.7211, -0.5289,  1.8362, -1.4268, -1.1033,  0.0696,
```

```
        0.5678,  0.5952,  0.2172,  0.5269,  1.4032, -0.3520, -0.7009,  0.0710,
       -0.2730, -1.4919, -1.3549,  0.1566, -1.0187,  0.0810])
tensor([-0.0040, -0.0026,  0.5494, -0.0119,  1.0810,  0.0770,  0.5700, -0.0109,
        0.6954, -0.0036, -0.0072, -0.0053,  1.8362, -0.0143, -0.0110,  0.0696,
        0.5678,  0.5952,  0.2172,  0.5269,  1.4032, -0.0035, -0.0070,  0.0710,
       -0.0027, -0.0149, -0.0135,  0.1566, -0.0102,  0.0810])
```

秘籍 4-2　激活函数可视化

问题

如何可视化激活函数？激活函数的可视化对于正确构建神经网络模型非常重要。通过可视化，您可以直观地了解激活函数的特性、变化以及对输入数据的影响。

解决方案

激活函数在神经网络中起着至关重要的作用，将数据从一层传输至另一层。为了直观地展示这一过程，可以将经过转换的数据以图表的形式呈现，并与实际的张量进行对比，从而实现激活函数的可视化。以下是采取的具体实现方法：
- 首先，选择一个样本张量。
- 将其转换为 PyTorch 变量。
- 应用所选的激活函数，得到另一个张量。
- 使用 Matplotlib 绘制出原始张量和转换后的张量。

编程实战

选择合适的激活函数不仅可以提高准确性，还有助于提取有意义的信息。

In [15]:
```
Import torch.nn.functional as F
from torch.autograd import Variable
import matplotlib.pyplot as plt
%matplotlib inline
```

In[16]:
```
x = torch.linspace(-10, 10, 1500)
x = Variable(x)
x_1 = x.data.numpy()      # 转换为 numpy 类型
```
In[17]:
```
y_relu = F.relu(x).data.numpy()
y_sigmoid = torch.sigmoid(x).data.numpy()
y_tanh = torch.tanh(x).data.numpy()
y_softplus = F.softplus(x).data.numpy()
```

在本脚本中，我们构建了一个线性空间中的数组，其范围介于 –10~10 之间，包含 1500 个样本点。随后，将此向量转换为 Torch 变量，再转换为 NumPy 变量，以便于绘制图形。接着，针对不同的激活函数进行计算。图 4-1~ 图 4-4 展示了这些激活函数的具体形式。

In[18]:
```
Plt.figure(figsize=(7, 4))
plt.plot(x_1, y_relu, c='blue', label='ReLU')
plt.ylim((-1, 11))
plt.legend(loc='best')
```

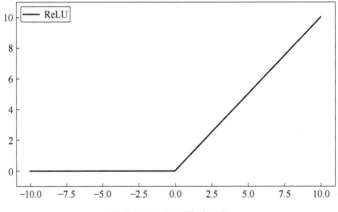

图 4-1　ReLU 激活函数

In[19]:
```
plt.figure(figsize=(7, 4))
plt.plot(x_1, y_sigmoid, c='blue', label='sigmoid')
plt.ylim((-0.2, 1.2))
plt.legend(loc='best')
```

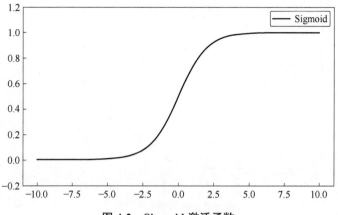

图 4-2　Sigmoid 激活函数

In [20]:
```
plt.figure(figsize=(7, 4))
plt.plot(x_1, y_tanh, c='blue', label='tanh')
plt.ylim((-1.2, 1.2))
plt.legend(loc='best')
```

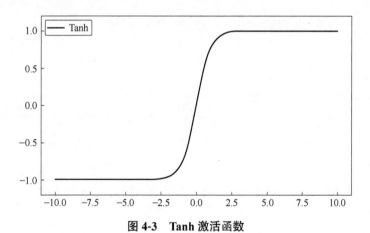

图 4-3　Tanh 激活函数

In [21]:
```
plt.figure(figsize=(7, 4))
plt.plot(x_1, y_softplus, c='blue', label='softplus')
plt.ylim((-0.2, 11))
plt.legend(loc='best')
```

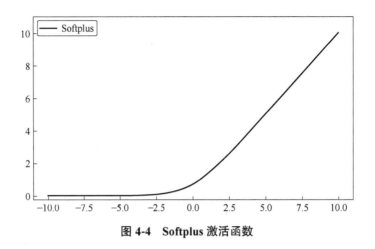

图 4-4　Softplus 激活函数

秘籍 4-3　基本的神经网络模型

问题

如何使用 PyTorch 构建一个基本的神经网络模型？

解决方案

在 PyTorch 中，构建基本的神经网络模型通常需要以下六个步骤：

- **准备训练数据**：收集和准备用于训练的数据集，包括输入特征和对应的目标标签。
- **创建基本神经网络模型**：使用 torch.nn 模块定义一个简单的神经网络模型，包括输入层、隐藏层和输出层。
- **初始化权重**：为模型的权重和偏置项进行初始化，以便开始训练。
- **计算损失函数**：选择适当的损失函数来衡量模型预测与实际标签之间的差异。
- **选择学习率**：根据问题的特点选择合适的学习率，以控制模型参数的更新速度。
- **优化损失函数**：使用优化算法（如随机梯度下降）来最小化损失函数，更新模型的参数。

编程实战

接下来，将按步骤来创建一个基本的神经网络模型。

In[22]:
```
Def prep_data():
    train_X = np.asarray([13.3,14.4,15.5,16.71,16.93,14.168,19.779,16.182,
                17.59,12.167,17.042,10.791,15.313,17.997,15.654,
                19.27,13.1])
    train_Y = np.asarray([11.7,12.76,12.09,13.19,11.694,11.573,13.366,
    12.596, 12.53,11.221,12.827,13.465,11.65,12.904,12.42,12.94,11.3])
    dtype = torch.FloatTensor
    X = Variable(torch.from_numpy(train_X).type(dtype),
            requires_grad=False).view(17,1)
    y = Variable(torch.from_numpy(train_Y).type(dtype),requires_grad=False)
    return X,y
```

为了展示一个示例神经网络模型，需要准备相应数据集并将数据类型转换为浮点张量。在项目开发过程中，构建数据集是一项独立的工作任务，数据准备应该以适当的方式进行。在上述代码中，train_X 和 train_Y 分别作为两个 NumPy 向量创建。随后，需将数据类型转换为浮点张量，以便在执行矩阵乘法运算时能够使用该数据。接着，将数据转换为 Variable 数据类型，由于 Variable 变量具有三个属性，有利于后续对对象进行微调。数据集中包含 17 个一维数据点。

Variable 类具有三个主要属性，即 data、grad 和 requires_grad，这些属性有助于对对象进行微调。

- data：保存了 Variable 中的张量数据。
- grad：保存了关于 Variable 的梯度。在反向传播期间，梯度信息会根据计算图自动累积到 grad 属性中。
- requires_grad：表示是否需要计算该 Variable 的梯度。如果设置为 True，则计算图会跟踪对该 Variable 的操作并计算梯度，否则不计算梯度。

In[23]:
```
# 获取动态参数
```

In[24]:
```
def set_weights():
    w = Variable(torch.randn(1),requires_grad = True)
    b = Variable(torch.randn(1),requires_grad=True)
    return w,b
```
In[25]:
```
# 部署神经网络模型
```
In[26]:
```
def build_network(x):
    y_pred = torch.matmul(x,w)+b
    return y_pred
```
In[27]:
```
# 在PyTorch中实现
import torch.nn as nn
f = nn.Linear(17,1) # 相比之下,在PyTorch中实现要简单一些。
F
```
Out[27]:
```
Linear(in_features=17, out_features=1, bias=True)
```

在上述代码中,函数set_weight()的作用是初始化神经网络模型在前向传播过程中所使用的随机权重。构建模型所需的两个张量分别为权重和偏置项。函数build_network()将权重与输入数据相乘,再加上偏置项,进而生成预测值。这是一个我们自定义的函数。若在PyTorch中实现相似功能,采用nn.Linear()模块会更加简洁,尤其在执行线性回归任务时。

In[28]:
```
# 计算损失函数
```
In[29]:
```
def loss_calc(y,y_pred):
    loss = (y_pred-y).pow(2).sum()
    for param in [w,b]:
        if not param.grad is None: param.grad.data.zero_()
    loss.backward()
    return loss.data
```
In[30]:
```
# 优化结果
```

In[31]:
```python
def optimize(learning_rate):
    w.data -= learning_rate * w.grad.data
    b.data -= learning_rate * b.grad.data
```

In[32]:
```python
learning_rate = 1e-4
```

In[33]:
```python
x,y = prep_data()      # x 为训练数据，y 为目标变量
w,b = set_weights()    # w, b 为参数
for i in range(5000):
    y_pred = build_network(x)    # 计算 wx + b 的函数
    loss = loss_calc(y,y_pred)    # 误差计算
    if i % 1000 == 0:
        print(loss)
    optimize(learning_rate)    # 关于 w, b 最小化损失函数值
```
tensor(5954.0488)
tensor(44.9320)
tensor(39.5382)
tensor(34.9094)
tensor(30.9371)

在确定网络结构后，为实现预测结果与实际输出的比对，以评估预测步骤的准确性，采用损失函数作为衡量系统准确性的标准，期望其值最小化。损失函数可能呈现不同的函数形态。那么，如何精确地确定损失函数在哪个迭代过程中达到最小值，即如何确定提供最优结果的迭代次数呢？为解决此问题，需要将优化算法应用于损失函数，以寻找最小的损失值。随后，可以提取对应于该迭代的参数。图 4-5 所示为真实值和预测张量。

In[34]:
```python
import matplotlib.pyplot as plt
%matplotlib inline
```

In[35]:
```python
x_numpy = x.data.numpy()
y_numpy = y.data.numpy()
```

```
y_pred = y_pred.data.numpy()
plt.plot(x_numpy,y_numpy,'o')
plt.plot(x_numpy,y_pred,'-')
```

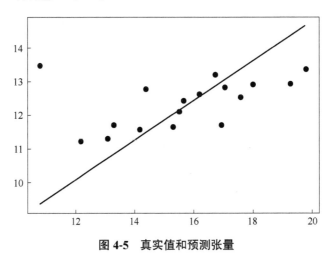

图 4-5　真实值和预测张量

秘籍 4-4　张量微分

问题

什么是张量微分，它与使用 PyTorch 框架执行计算图有什么关系？

解决方案

计算图网络是由节点构成的，并通过函数进行连接。该网络包含两种不同类型的节点：依赖节点与独立节点。依赖节点需等待其他节点输出结果后，方能处理输入。而独立节点之间相互连接，可为常量或代表计算结果。张量微分是在计算图环境中实施高效计算的一种方法。

编程实战

在计算图中，张量微分展现出了高效的特点，这得益于张量可以被视作并行节点、多进程节点或多线程节点进行计算。因此，大多数深度学习和神经计算框架皆具备此类张量微分功能。

Autograd 是一种应用于张量微分的函数，其主要功能在于计算误差函数的梯度或导数，进而实现神经网络中权重与偏置的微调。在训练过程中，Autograd 通过设置学习率及迭代次数，力求降低误差值或损失函数值。

在运用张量微分的过程中，需借助 nn.backward( ) 函数。下面将通过实例演示误差梯度是如何进行反向传播的。为了绘制损失函数的曲线更新，或寻找损失函数最小值及其所在位置和移动方向，需进行导数计算。张量微分便是计算图中计算函数斜率的一种方法。

In[36]:

x = Variable(torch.ones(4, 4) * 12.5, requires_grad=True)

In[37]:

x

Out[37]:

tensor([[12.5000, 12.5000, 12.5000, 12.5000], [12.5000, 12.5000, 12.5000, 12.5000], [12.5000, 12.5000, 12.5000, 12.5000], [12.5000, 12.5000, 12.5000, 12.5000]], requires_grad=True)

In[38]:

fn = 2 * (x * x) + 5 * x + 6

2x^2 + 5x + 6

In[39]:

fn.backward(torch.ones(4,4))

In[40]:

print(x.grad)
tensor([[55., 55., 55., 55.],
        [55., 55., 55., 55.],
        [55., 55., 55., 55.],
        [55., 55., 55., 55.]])

在上述代码脚本中，x 是一个样本张量，须执行自动梯度计算，因此 requires_grad 参数设置为 True。变量 fn 是由 x 变量创建的线性函数。通过应用 backward( ) 函数，实现反向传播计算。最后输出 .grad 成员变量中保存的张量微分结果。

小结

本章讨论了各类激活函数及其在不同场景下的应用方法。挑选最优激活函数的策略或系统应以准确性原则为基础，应始终动态灵活地选择能够产生最佳结果的激活函数。本章中，我们利用小样本张量构建了基础的神经网络模型，通过权重优化更新，最终得出预测结果。在后续章节中，将展示更多实例。

第 5 章

PyTorch 中的监督学习

监督学习作为机器学习领域中最复杂的分支之一，其在各个领域，如人工智能、认知计算和自然语言处理等方面具有广泛应用。总的来说，与机器学习相关的文献可分为三类：监督学习、无监督学习和强化学习。在监督学习中，算法学习如何对模型的输出进行识别；因此，它是任务驱动的，任务可以是分类或回归。在无监督学习中，算法从数据中学习数据模式；因此，它能够泛化新的数据集，并通过评估一组输入特征来进行学习。而强化学习则是依据系统对情境的反应进行学习的过程。

本章将详细介绍回归技术，并结合机器学习方法解释回归方法的输出在业务场景中的意义。机器学习算法分类如图 5-1 所示。

分类任务（Classification）：在数据库中，若每个对象或行代表一个事件，我们的任务便是将这些事件划分至不同的类别之中。将记录分配至相应类别的过程称为分类，其中目标变量的特定标签或标记与事件相对应。以银行数据库为例，客户可被标记为忠实客户或非忠实客户；在医疗记录数据库中，患者则根据其疾病类别进行标记；而在电信行业，订阅者可以被划分为流失或非流失客户。以上皆为监督算法执行分类的实例。分类一词源于目标列中呈现的类别。

回归任务（Regression）：在回归学习的过程中，核心目标在于预测连续性变量的数值。例如，在具备一套房产的各项属性数据，如卧室数量、面积、周边环境、乡镇等因素的情况下，可以推算出房产售价。此类情况适合采用回归模型进行处理。其他相关应用还包括预测股票价格或企业销售额、收入及利润等。

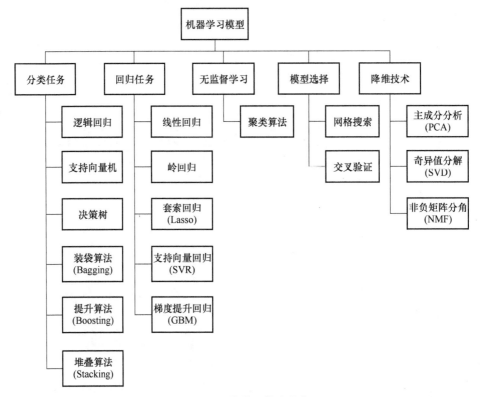

图 5-1 机器学习算法分类

在**无监督学习**（Unsupervised Learning）算法中，我们旨在探究数据集中观测、记录或行的自然分组，并无结果变量、标记或标签。自然分组应满足以下要求：组内相似性最大化，组间相似性最小化。

在现实世界的场景中，有时回归分析并不能对目标变量的预测提供帮助。在监督式回归技术中，输入数据亦称为训练数据。针对每个记录，都对应一个连续数值的标签。通过训练过程，模型能够预测正确的输出，此过程将持续进行，直至达到所需的准确度水平。为解析数据集中所存在的模式，需要运用高级回归方法。

线性回归介绍

线性回归分析被认为是众多统计技术中最为可靠、应用最为便捷且广泛的方法之一。这一方法基于假设因变量与自变量之间存在线性且可加的关系。线性回归的核心目标在于利用自变量对因变量或目标变量进行预测。线性回归模型的构建公式为

$$Y = \alpha + \beta X$$

此公式具备一个特性，即对于每一个 X 变量，Y 的预测为一个直线函数，在保持其他所有变量不变的前提下，不同 X 变量对预测的贡献是可加的。它们与 Y 各自的线性关系的斜率即为变量系数。这些系数与截距是通过最小二乘法估算得到的（即将其设定为使模型在数据样本内拟合的平方误差之和最小的唯一值）。

模型的预测误差通常被认为服从独立同分布的正态分布。当 β 系数为零时，输入变量 X 对因变量无影响。普通最小二乘（Ordinary Least Square，OLS）法的目标是最小化残差的平方和。残差是指回归线上的点与散点图中实际数据点之间的差值。这一过程旨在估算多元线性回归模型中的 β 系数。

在一个包含 15 个人的样本数据集中，我们观察到每个人的身高和体重均有记录。在这样的情况下，仅依据身高数据，我们能够通过线性回归技术预测一个人的体重吗？答案基本上是肯定的。

人（序号）	1	2	3	4	5	6	7	8	9	10	11	12	13	14	15
身高	58	59	60	61	62	63	64	65	66	67	68	69	70	71	72
体重	115	117	120	123	126	129	132	135	139	142	146	150	154	159	164

为了图形化展示，用 x 轴表示身高，用 y 轴表示体重。线性回归方程如图 5-2 所示，其中截距为 87.517，斜率系数为 3.45。数据点以点的形式呈现，连线为线性回归预测。

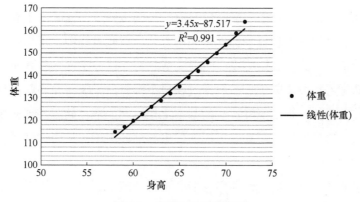

图 5-2　身高与体重的关系

事实上，现实生活中大部分情况都表现出与线性关系不同的其他形式的关系，那么为何我们倾向于假定因变量与自变量之间存在线性关系呢？以下内容将探讨在此背景下依然坚持使用线性关系的原因。

首先，线性关系易于理解和解释。其次，可以通过转换现有的非线性关系，使其变为线性关系。最后，线性关系易于生成预测。

预测建模领域主要关注于在预测模型中最小化误差或进行尽可能准确的预测。线性回归作为一种统计学方法，原本用于解析数值变量间输入与输出关系而设计，后被机器学习领域借鉴。既作为一种统计算法，也作为一种机器学习算法。线性回归模型的建立依赖于以下假设：

- **线性关系**：线性回归假设因变量与自变量之间具有线性关系。在这种情况下，回归方程中的因变量变化与自变量变化呈比例关系。
- **无多重共线性**：如果输入特征空间中有多于两个自变量，那么自变量之间不应该存在多重共线性，即输入特征不应该相关。
- **无自相关性**：线性回归模型假设误差项（残差）之间没有自相关。
- **同方差性**：线性回归模型假设误差项的方差在不同自变量取值下保持不变，即误差方差应该恒定。
- **误差项正态分布**：误差项基本上定义为实际观测值与预测值之间的差异，它应该服从正态分布。

线性回归方法具有多种变体，但在机器学习领域，将其视为同一方法的不同表现。以单一自变量预测因变量为例，称之为简单线性回归模型。当自变量为多个时，相应的模型则被称为多元线性回归模型。普通最小二乘法是一种用于预测线性回归模型的统计方法，因此有时线性回归模型也被称为普通最小二乘模型。

线性回归分析对**缺失值**和**异常值**具有较高的敏感度，这是由于其统计计算方法依赖于变量之间的平均值、标准差和协方差。在此过程中，平均值对异常值的变动尤为敏感。因此，在构建线性回归模型之前，有必要对异常值进行剔除。

在机器学习相关文献中，一般通过**梯度下降算法**来获得在回归模型中最小化误差的最佳 β 系数。梯度下降算法的基本工作原理如下：始于一个初始值，最好是从零开始，通过设定学习率，定期迭代更新缩放因子，从而达到最小化误差项的目的。

基于机器学习方法的线性回归分析需进行特定数据预处理，以确保原始数据完整性。数据转换则是提高模型稳健性的关键步骤。

秘籍 5-1 监督模型的数据准备

问题

如何使用 PyTorch 为创建监督学习模型执行数据准备？

解决方案

本节将使用开源数据集 mtcars.csv，探讨如何构建输入和输出张量。该数据集为一个回归数据集，旨在研究不同汽车属性与价格之间的关系。

编程实战

首先，需要导入必要的库。

```
In[1]:
import torch
import pandas as pd
import numpy as np
import matplotlib.pyplot as plt
from torch.autograd import Variable
import torch.nn.functional as F
%matplotlib inline
In[2]:
torch.__version__
Out[2]:
1.12.1+cu113
In[3]:
df = pd.read_csv("https://raw.githubusercontent.com/pradmishra1/
PublicDatasets/main/mtcars.csv")
```

In[4]:
```
del df['Unnamed: 0']
df.head()
```

Model	MPG	Cyl	Disp	HP	Drat	Wt	Qsec	Vs	Am	Gear	Carb
0	Mazda RX4	21.0	6	160.0	110	3.90	2.620	16.46	0	1	4	4
1	Mazda RX4 Wag	21.0	6	160.0	110	3.90	2.875	17.02	0	1	4	4
2	Datsun 710	22.8	4	108.0	93	3.85	2.320	18.61	1	1	4	1
3	Hornet 4 Drive	21.4	6	258.0	110	3.08	3.215	19.44	1	0	3	1
4	Hornet Sportabout	18.7	8	360.0	175	3.15	3.440	17.02	0	0	3	2

在上述示例中，监督算法的预测变量为 Qsec，其作用在于预测汽车每加仑的行驶里程。在本例中，数据类型的转换十分重要。首先，需要将采用 NumPy 格式存储的数据导入至 PyTorch 张量格式中。默认情况下，张量格式为浮点数。若在优化函数时使用浮点数格式，可能导致错误发生，因此调整张量数据类型显得至关重要。可以通过 unsqueeze 函数重新定义张量的数据类型，并设定维度为 1。

In[5]:
```
torch.manual_seed(1234)      # 为了可复现
```

In[6]:
```
x = torch.unsqueeze(torch.from_numpy(np.array(df.qsec)),dim=1)
y = torch.unsqueeze(torch.from_numpy(np.array(df.mpg)),dim=1)
```

In[7]:
```
x[0:10]
```

Out[7]:
tensor([[16.4600], [17.0200], [18.6100], [19.4400], [17.0200], [20.2200], [15.8400], [20.0000], [22.9000], [18.3000]], dtype=torch.float64)

In[8]:
```
y[0:10]
```

Out[8]:
tensor([[21.0000], [21.0000], [22.8000], [21.4000], [18.7000], [18.1000], [14.3000], [24.4000], [22.8000], [19.2000]], dtype=torch.float64)

为了复现相同的结果，需设定手动种子，此处采用 torch.manual_seed（1234）。虽然数据类型看似为张量，但若运用 type 函数进行检测，则会显示为 double，这是因为优化函数需要 double 类型的张量。

秘籍 5-2　前向和反向传播神经网络

问题

如何构建一个基于 Torch 的神经网络类函数,以构建前向传播方法?

解决方案

设计神经网络类函数,包括从输入层至隐藏层的处理(前向传播)以及从隐藏层至输出层的处理(反向传播)。在神经网络结构中,还需设定隐藏层内神经元的数量。

编程实战

在 Net( ) 类中,首要步骤是初始化特征层、隐藏层和输出层。接下来,采用修正线性单元(ReLU)作为隐藏层的激活函数,并引入反向传播算法。

In[9]:
```
class Net(torch.nn.Module):
    def __init__(self, n_feature, n_hidden, n_output):
        super(Net, self).__init__()
        self.hidden = torch.nn.Linear(n_feature, n_hidden)   #神经网络的隐
                                                              藏层
        self.predict = torch.nn.Linear(n_hidden, n_output)   #输出层
    def forward(self, x):
        x = F.relu(self.hidden(x))      #隐藏层的激活函数
        x = self.predict(x)             #线性输出
        return x
```

图 5-3 展示了 ReLU 激活函数。这种激活函数在多种神经网络模型中具有广

泛的应用。然而，选择激活函数时，应以准确性为依据。若采用其他激活函数（如 Sigmoid 函数）能提升准确性，则应予以考虑。

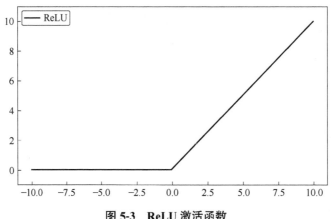

图 5-3　ReLU 激活函数

在当前的示例中，我们所采用的网络架构是基于监督学习模型的。其中，n_feature 代表了输入层神经元的数量。由于我们仅有一个输入变量 Qsec，因此将 n_feature 设为 1。隐藏层神经元的数量则可根据输入及学习模型所要求的准确度进行调整。在本例中，设定 n_hidden 为 20，这意味着隐藏层 1 含有 20 个神经元，而输出层神经元数量为 1。

In[10]:
```
net = Net(n_feature=1, n_hidden=20, n_output=1)
net.double()
print(net)  # 神经网络架构
Net(
  (hidden): Linear(in_features=1, out_features=20, bias=True)
  (predict): Linear(in_features=20, out_features=1, bias=True)
)
```

In[11]:
```
optimizer = torch.optim.SGD(net.parameters(), lr=0.2)
loss_func = torch.nn.MSELoss()
# 用于回归均方损失
```

这段代码是一个简单的神经网络示例，用于回归任务。
- **定义神经网络架构**：
  - Net 类定义了一个神经网络，具有以下结构：

- 输入层：1 个特征（n_feature=1）。
- 隐藏层：20 个神经元（n_hidden=20）。
- 输出层：1 个输出（n_output=1）。
- net.double( ) 将网络的参数转换为双精度浮点数。
- 初始化优化器和损失函数：
  - optimizer 使用随机梯度下降（SGD）算法来更新网络的参数，学习率为 0.2。
  - loss_func 是均方误差（MSE）损失函数，用于回归任务。

优化函数的主要任务是最小化与参数及学习率相关的损失函数。在机器学习领域，优化函数旨在训练过程中寻找最优的参数组合，从而使模型在给定数据上展现出最佳性能。学习率则决定了每次参数更新的步长，它影响着模型在每次迭代中对梯度的权衡程度。选用适当的优化函数和学习率，有助于提高模型训练速度，并使其更有效地逼近最优解。在此示例中，设定学习率为 0.2，并将神经网络的参数传递给优化器。PyTorch 提供了多种不同的优化函数，以满足不同的需求：

- **SGD**：实现随机梯度下降（可选带有动量）。参数包括动量、学习率和权重衰减。
- **Adadelta**：自适应学习率。具有五个不同的参数：网络参数、用于计算平方梯度滑动平均的系数、用于实现模型数值稳定性的参数项、学习率和权重衰减参数以应用正则化。
- **Adagrad**：用于在线学习和随机优化的自适应次梯度方法。参数包含用于优化的参数的迭代对象，具有权重衰减的学习率和学习率衰减。
- **Adam**：一种用于随机优化的方法。该函数具有六个不同的参数：用于优化的参数迭代对象、学习率、beta（用于计算梯度及其平方的滑动平均的系数）、用于提高数值稳定性的参数等。
- **ASGD**：通过平均的方法加速随机近似。该函数具有五个不同的参数：用于优化的参数迭代对象、学习率、衰减项、权重衰减等。
- **RMSprop** 算法：使用计算的梯度的大小来规范化梯度。
- **SparseAdam**：实现适用于稀疏张量的 Adam 算法的延迟版本。在这个变种中，只有梯度中出现的矩阵才会被更新，并且只有梯度中的那部分会应用于参数。

在运行监督学习模型之前，除优化函数外，还需选择一个适当的损失函数。在 PyTorch 库中，提供了众多不同的损失函数以供选择。

- MSELoss：创建一个衡量输入变量和目标变量之间均方误差的准则。对于回归相关的问题，这是最佳的损失函数。

In[12]:

optimizer

Out[12]:

SGD (Parameter Group 0 dampening: 0 foreach: None lr: 0.2 maximize: False momentum: 0 nesterov: False weight_decay: 0)

In[13]:

loss_func

Out[13]:

MSELoss()

In[14]:

\# 开启交互模式

plt.ion()

在运行监督学习中的回归模型之后，需要输出实际值与预测值，并以图形形式展示它们。因此，需启用 Matplotlib 的交互模式。在交互模式下，图表将自动更新，无需手动调用 plt.show( )。

秘籍 5-3　优化和梯度计算

问题

如何使用 PyTorch 构建一个基本的监督式神经网络训练模型，并进行不同的迭代？

解决方案

在 PyTorch 中，构建基本神经网络模型涉及以下六个关键步骤：准备训练数据、初始化权重、构建基本网络结构、确定损失函数、设定学习率以及采用优化算法最小化损失函数相对于模型参数的梯度。

编程实战

接下来,将逐步创建一个基本的神经网络模型。

In[15]:
```
for t in range(100):
    prediction = net(x)         # 输入 x 并根据 x 进行预测
    loss = loss_func(prediction, y)    # 参数必须是(1.nn 输出,2.目标标签)
    optimizer.zero_grad()       # 清除此训练步骤的梯度
    loss.backward()             # 反向传播,计算梯度
    optimizer.step()            # 通过梯度下降执行一步参数更新
    if t % 50 == 0:
        # 绘制并展示参数学习过程
        plt.cla()
        plt.scatter(x.data.numpy(), y.data.numpy())
        plt.plot(x.data.numpy(), prediction.data.numpy(), 'g-', lw=3)
        plt.text(0.5, 0, 'Loss=%.4f' % loss.data.numpy())
        plt.show()
plt.ioff()
```

图 5-4 展示了模型在首次迭代与最后迭代中的最终预测成果。在初始阶段,损失函数值为 276.9186。经过优化调整后,损失函数下降至 35.1890。图中展示了拟合的回归线及其与数据集的契合程度。

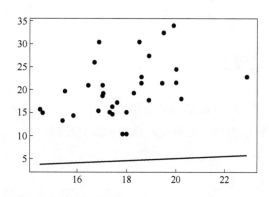

损失函数值=276.9186

图 5-4 模型优化前后拟合回归线对比

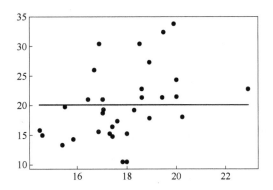

损失函数值=35.1890

图 5-4　模型优化前后拟合回归线对比（续）

秘籍 5-4　查看预测结果

问题

如何从基于 PyTorch 的监督学习模型中提取最佳结果？

解决方案

计算图网络是由节点构成，并通过函数进行连接的一种模型。我们可以运用多种方法来最小化误差函数，从而获得最优的预测模型。这些方法包括增加迭代次数、估算损失函数、优化函数、输出实测值与预测值，并在图中进行展示。

编程实战

在应用张量微分的过程中，需借助 nn.backward( ) 方法。下面通过实例演示误差梯度在神经网络中的反向传播过程。grad 变量用于保存张量微分的最终结果。训练过程的可视化如图 5-5 所示。

In[16]:
```
optimizer = torch.optim.SGD(net.parameters(), lr=0.001)
loss_func = torch.nn.MSELoss()    #用于回归均方损失
```

In [17]:
```
for t in range(1000):
    prediction = net(x)                       # 输入x并根据x进行预测
    loss = loss_func(prediction, y)           # 参数必须是(1.nn输出, 2.目标标签)
    optimizer.zero_grad()                     # 清除此训练步骤的梯度
    loss.backward()                           # 反向传播，计算梯度
    optimizer.step()                          # 通过梯度下降执行一步参数更新

    if t % 100 == 0:
        # 绘制并展示训练过程
        plt.cla()
        plt.scatter(x.data.numpy(), y.data.numpy())
        plt.plot(x.data.numpy(), prediction.data.numpy(), 'g-', lw=3)
        plt.text(0.5, 0, 'Loss=%.4f' % loss.data.numpy())
        plt.show()
plt.ioff()    # 关闭交互模式
```

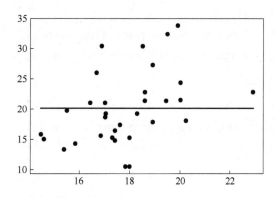

损失函数值=35.1890

图 5-5　训练过程的可视化图

我们可以通过调整以下参数来提高回归用例的监督学习模型的准确性：
- 迭代次数：增加迭代次数可以提高模型的拟合效果。
- 损失函数类型：选择适合回归问题的损失函数，如均方误差（MSE）。
- 优化算法的选择：选择合适的优化算法，如随机梯度下降（SGD）。
- 学习率：调整学习率可以影响模型的收敛速度。
- 学习率衰减：逐步减小学习率有助于更稳定地优化模型。
- 动量：在优化中使用动量可以加速收敛过程。

真实数据集的样子如下所示：

In[18]:

df.head()

Model	MPG	Cyl	Disp	HP	Drat	Wt	Qsec	Vs	Am	Gear	Carb
0	Mazda RX4	21.0	6	160.0	110	3.90	2.620	16.46	0	1	4	4
1	Mazda RX4 Wag	21.0	6	160.0	110	3.90	2.875	17.02	0	1	4	4
2	Datsun 710	22.8	4	108.0	93	3.85	2.320	18.61	1	1	4	1
3	Hornet 4 Drive	21.4	6	258.0	110	3.08	3.215	19.44	1	0	3	1
4	Hornet Sportabout	18.7	8	360.0	175	3.15	3.440	17.02	0	0	3	2

以下脚本演示了从 mtcars.csv 数据集中读取 MPG 列和 Qsec 列的方法（注：第一列数字为数据的行编号）。通过运用 unsqueeze 函数，将这两个变量转换为张量，并将其应用于神经网络模型进行预测。

In[19]:
```
x = torch.unsqueeze(torch.from_numpy(np.array(df.mpg)),dim=1)
y = torch.unsqueeze(torch.from_numpy(np.array(df.qsec)),dim=1)
```

In[20]:
```
optimizer = torch.optim.SGD(net.parameters(), lr=0.2)
loss_func = torch.nn.MSELoss()     #用于回归均方损失
```

In[21]:
```
plt.ion()      #开启交互模式
```

In[22]:
```
for t in range(1000):
    prediction = net(x)                      #输入x并根据x进行预测
    loss = loss_func(prediction, y)          #参数必须是(1.nn输出,2.目标标签)
    optimizer.zero_grad()                    #清除此训练步骤的梯度
    loss.backward()                          #反向传播,计算梯度
    optimizer.step()                         #通过梯度下降执行一步参数更新
    if t % 200 == 0:
        #绘制并展示训练过程
```

```
        plt.cla()
        plt.scatter(x.data.numpy(), y.data.numpy())
        plt.plot(x.data.numpy(), prediction.data.numpy(), 'g-', lw=3)
        plt.text(0.5, 0, 'Loss=%.4f' % loss.data.numpy())
        plt.show()
plt.ioff()        #关闭交互模式
```

经过1000次迭代，模型收敛（见图5-6）。

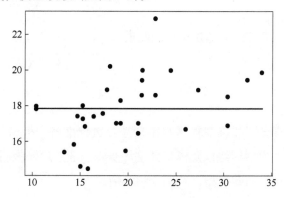

损失函数值=3.0934

图 5-6　经过 1000 次迭代，模型收敛

在 Torch 库中，神经网络的构建与 nn 模块密切相关。接下来，将详细探讨这个模块。

在 PyTorch 中，**torch.nn** 包是用于构建神经网络的基本工具，它提供了构建、训练和评估神经网络所需的大部分功能，主要包括以下模块：

- 线性层：nn.Linear、nn.Bilinear。
- 卷积层：nn.Conv1d、nn.Conv2d、nn.Conv3d、nn.ConvTranspose2d。
- 非线性函数：nn.Sigmoid、nn.Tanh、nn.ReLU、nn.LeakyReLU。
- 池化层：nn.MaxPool1d、nn.AveragePool2d。
- 循环网络：nn.LSTM、nn.GRU。
- 归一化：nn.BatchNorm2d。
- **Dropout**：nn.Dropout、nn.Dropout2d。
- 嵌入层：nn.Embedding。
- 损失函数：nn.MSELoss、nn.CrossEntropyLoss、nn.NLLLoss。

标准分类算法属于监督学习算法的一种变体，在此算法中，目标变量为一个类别变量，而特征则可能包括数值型和分类型。

秘籍 5-5　监督模型逻辑回归

问题

如何使用 PyTorch 部署逻辑回归模型？

解决方案

使用 PyTorch 部署逻辑回归模型涉及数据准备、模型构建、损失函数定义、优化器选择、模型训练以及模型评估等步骤。通过逐步执行这些步骤，可以构建和部署一个逻辑回归模型，并用于预测新样本。

编程实战

首先，需要导入 PyTorch 相关的库，包括 torch、torch.nn 和 torch.optim。

In [23]:
```
import torch
from torch.autograd import Variable
import torch.nn as nn
import torch.nn.functional as F
import matplotlib.pyplot as plt
import torch.optim as optim
```
torch.manual_seed(1)

接下来进行逻辑回归模型的数据准备过程。

In [24]:
```
# 逻辑回归的数据准备
n_data = torch.ones(100,2)
x0 = torch.normal(2*n_data,1)
y0 = torch.zeros(100)
```

In[25]:
x1 = torch.normal(-2*n_data,1)
y1 = torch.ones(100)

In[26]:
x = torch.cat((x0,x1),0).type(torch.FloatTensor)
y = torch.cat((y0,y1),).type(torch.LongTensor)

In[27]:
变量转换
x, y = Variable(x), Variable(y)

In[28]:
逻辑回归模型的样本数据准备

接下来看一下用于分类的样例数据集。图 5-7 所示为逻辑回归的样例数据。

In[29]:
plt.scatter(x.data.numpy()[:,0], x.data.numpy()[:,1],c=y.data.numpy(),s=100)
plt.show()

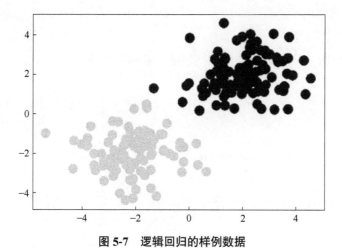

图 5-7 逻辑回归的样例数据

接下来,为逻辑回归模型设置神经网络模块。

In[30]:
class Net(torch.nn.Module):
    def __init__(self, n_feature, n_hidden, n_output):
        super(Net, self).__init__()

```python
        self.hidden = torch.nn.Linear(n_feature, n_hidden)    # 隐藏层
        self.out = torch.nn.Linear(n_hidden, n_output)         # 输出层
    def forward(self, x):
        x = F.sigmoid(self.hidden(x))    # 隐藏层的激活函数
        x = self.out(x)                  # 线性输出
        return x
```

下一步，配置并查看神经网络的参数。

In [31]:
```
net = Net(n_feature=2,n_hidden=10,n_output=2)
print(net)
Net(
  (hidden): Linear(in_features=2, out_features=10, bias=True)
  (out): Linear(in_features=10, out_features=2, bias=True)
)
```

In [32]:
```
# 损失与优化器
# softmax 在内部进行计算
# 设置要更新的参数
```

最后，执行迭代，并找到图例问题的最佳解决方案。

In [33]:
```
# net(x)
```

In [34]:
```
optimizer = torch.optim.SGD(net.parameters(),lr=0.02)
```

In [35]:
```
loss_func = torch.nn.CrossEntropyLoss()
```

In [36]:
```
plt.ion()   # 开启交互模式
```

In [37]:
```
for t in range(100):
    out = net(x)           # 输入 x 并根据 x 进行预测
    loss = loss_func(out, y)      # 参数必须是(1.nn 输出, 2.目标标签)
    optimizer.zero_grad()         # 清除此训练步骤的梯度
```

```
    loss.backward()          # 反向传播,计算梯度
    optimizer.step()         # 通过梯度下降执行一步参数更新
    if t % 10 == 0 or t in [3,6]:
        # 绘制并展示参数学习过程
        plt.cla()
        _,prediction = torch.max(F.softmax(out,dim=1),1)
        pred_y = prediction.data.numpy().squeeze()
        target_y = y.data.numpy()
        plt.scatter(x.data.numpy()[:,0],
                    x.data.numpy()[:,1],
                    c = pred_y,s=100,lw=0)
        accuracy = sum(pred_y == target_y)/200.0
        plt.text(1.5, -4, 'Accuracy=%.2f' % accuracy)
        plt.show()
plt.ioff()   #关闭交互模式
```

在第一次迭代过程中,准确率几乎达到99%,随后模型在训练数据上达到100%的准确率(见图5-8和图5-9)。

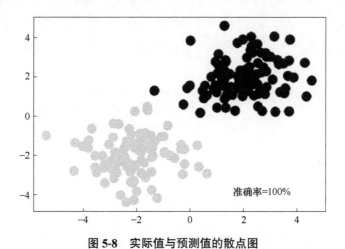

图5-8　实际值与预测值的散点图

最终准确率达到了100%,这显然是过拟合现象,然而,可通过引入丢弃率(Dropout rate)来调控过拟合,此方法在前述章节中已有详述。

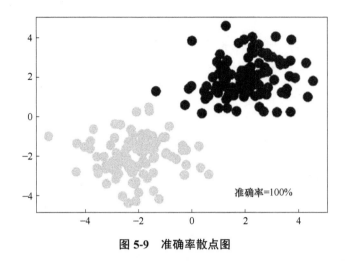

图 5-9　准确率散点图

小结

本章介绍了监督学习算法的两种主要类型,即线性回归与逻辑回归,并探讨了它们如何运用样本数据集以及在 PyTorch 框架中的实现过程。这两种算法均属于线性模型,其中,线性回归旨在预测实数值输出,而逻辑回归则旨在将某一类别与其他类别区分开来。虽然逻辑回归示例中仅考虑了二分类问题,但实则可以将其扩展至多类别分类模型。

第 6 章

使用 PyTorch 对深度学习模型进行微调

深度学习模型与生物神经元之间的连接方式和信息传递方式有着非常深厚的根源。

深度学习具有明确的应用场景，尤其在单一功能的传统机器学习技术难以应对现实生活中的挑战时。例如，在数据维度较大（高达数千维）的情况下，常规的机器学习算法通常难以对结果变量进行准确预测或分类。这些机器学习算法在计算效率上较差，资源消耗较大，且模型收敛难度较高。目标检测、图像分类以及图像分割等领域成为深度学习的突出应用实例。

最常用的深度学习算法可以分为三类：

- **卷积神经网络**：主要适用于高度稀疏的数据集、图像分类、图像识别、目标检测等。
- **循环神经网络**：适用于处理序列信息，特别是当数据生成方式中存在内部的顺序结构时。包括音乐、自然语言、音频和视频等，其信息存在时序性。
- **深度神经网络**：通常适用于单层机器学习算法无法正确分类或预测的情况。这类算法存在三种变体。
  - **宽网络**（Wide Network）：隐藏层中的神经元数量通常比前一层更多。这样的网络结构可以增加模型的表示能力，提高对复杂数据模式的学习能力。

- **深网络**（Deep Network）：隐藏层的数量比通常的神经网络模型更多。通过增加隐藏层的数量，网络可以学习到更多层次的特征表示，从而更好地捕捉数据中的抽象和复杂模式。
- **深而宽网络**（Deep and Wide Network）：是将宽网络和深网络的特点结合起来，既增加了隐藏层的数量，又增加了每层隐藏层中的神经元数量。这种网络结构可以同时具备较高的表示能力和学习能力，能够处理更加复杂和高维度的数据。

本章将探讨如何通过调整超参数来优化深度学习模型。需要注意的是，参数与超参数之间存在一定的区别。在深度学习模型中，我们关注的是如何调整超参数，因为它们对模型的性能有着至关重要的影响。与参数（权重，会随着初始值、学习率和迭代次数的变化而不断调整）不同，超参数是在模型构建过程中进行设置的，以实现最佳性能。正如第 3 章所阐述的，合理调整超参数对于深度学习模型的微调至关重要，从而获得理想的结果。

秘籍 6-1　构建顺序神经网络

问题

在 PyTorch 中，是否可以像在 Keras 中一样构建序贯神经网络模型，而不是声明整个神经网络模型？

解决方案

针对神经网络模型的构建，若逐行声明各项参数，如神经元数量、隐藏层数量、迭代次数、损失函数选择、优化函数选择、权重分布选择等，将会显得颇为繁琐，且在扩展模型时易于出现错误。为避免此类问题，可以采用高级函数来简化神经网络模型的构建，无需逐行声明所有参数。这样既减少了冗余操作，又允许后端设置一些默认参数，从而使用户仅需提供必要的超参数即可获得所需结果。确切地说，我们无需声明整个神经网络模型。

编程实战

接下来将探讨如何在 PyTorch 中创建此类模型。在 PyTorch 中,神经网络模块包含一个功能性 API,其中涵盖了前面章节中讨论过的各种激活函数。

In[1]:
```
import torch
import torch.nn.functional as F
```

在以下代码脚本中,我们创建了一个简单的神经网络模型,其中使用线性函数作为输入层到隐藏层和隐藏层到输出层的激活函数。

在下面的代码中,需要定义 Net 类、特征、隐藏神经元和激活函数。这些元素都可以通过顺序模块进行简便替换。

In[2]:
```
# 用简单的顺序网络替换以下代码中的 Net 类
class Net(torch.nn.Module):
    def __init__(self, n_feature, n_hidden, n_output):
        super(Net, self).__init__()
        self.hidden = torch.nn.Linear(n_feature, n_hidden)    # 隐藏层
        self.predict = torch.nn.Linear(n_hidden, n_output)    # 输出层
    def forward(self, x):
        x = F.relu(self.hidden(x))     # 隐藏层的激活函数
        x = self.predict(x)            # 线性输出
        return x
```

通过修改上述代码中的类函数并将其替换为 Sequential 函数,可以实现上述功能。Keras 函数具有替代 TensorFlow 函数的能力,这意味着仅需几行 Keras 代码即可替代大量 TensorFlow 代码。换句话说,Keras 提供了更高级的抽象和简洁的接口,使神经网络模型的构建和训练变得更加简单且易于理解。在 PyTorch 中同样可以实现类似功能,无需引入外部模块。

例如,在以下代码中,net2 为一个顺序模型,而 net1 调用前述脚本中定义的网络。从可读性角度审视,net2 相较于 net1 更具优势。

In[3]:
```
net1 = Net(1, 100, 1)
```

In[4]:
```
# 简单快捷的方式构建网络
net2 = torch.nn.Sequential(
    torch.nn.Linear(1, 100),
    torch.nn.ReLU(),
    torch.nn.Linear(100, 1)
)
```

如果输出 net1 和 net2 的模型架构，可以看到它们的结构是一样的。

In[5]:
```
print(net1)      # net1 架构
print(net2)      # net2 架构
Net(
  (hidden): Linear(in_features=1, out_features=100, bias=True)
  (predict): Linear(in_features=100, out_features=1, bias=True)
)
Sequential(
  (0): Linear(in_features=1, out_features=100, bias=True)
  (1): ReLU()
  (2): Linear(in_features=100, out_features=1, bias=True)
)
```

秘籍 6-2　确定批量的大小

问题

如何使用 PyTorch 对深度学习模型进行批量数据训练？

解决方案

深度学习模型的训练往往依赖于大量的数据标注。在此训练过程中，通常寻找一组最优的权重和偏差，以使损失函数相对于目标标签达到最小化。若训练过程能够有效地逼近该函数，则预测或分类的准确性将得以提高。

编程实战

训练深度学习网络有两种方法：批量训练和在线训练。训练方法的选择取决于所采用的学习算法。若使用反向传播算法，在线训练表现出较高的适用性。而对于具有多层反向传播与前向传播的**深而宽网络**模型，批量训练则更具优势。

In[6]:
```
import torch
import torch.utils.data as Data
torch.manual_seed(1234)     # 为了可复现
```

在训练时，将批次大小（Batch Size）设定为 5。然后，可以尝试将批次大小调整为 8，并观察相应的影响。在在线训练过程中，权重和偏差会根据预测结果与实际结果的差异，针对每个训练样本进行实时更新。而在批量训练过程中，损失函数通常是针对整个批次中的所有样本进行计算的，然后将得到的损失值作为整个批次的损失。这个损失值将用于优化模型参数，从而使模型更好地拟合训练数据。

In[7]:
```
BATCH_SIZE = 5
```
In[8]:
```
x = torch.linspace(1, 10, 10)        # 这是 x 数据（torch 张量）
y = torch.linspace(10, 1, 10)        # 这是 y 数据（torch 张量）
```
In[9]:
```
torch_dataset = Data.TensorDataset(x, y)
loader = Data.DataLoader(
    dataset=torch_dataset,      # torch TensorDataset 格式
    batch_size=BATCH_SIZE,      # 小批次的大小
    shuffle=True,               # 随机打乱以进行训练
    num_workers=2,              # 用于加载数据的子进程数
)
```

在进行五次数据集迭代训练后（这里省略了训练的代码），可以按批次和步长输出训练时用到的数据。若将在线训练与批量训练进行比较，则批量训练呈现

出更多的优势。当训练大规模数据集时，将面临内存限制的问题。在这种情况下，批量训练成为解决之道，助力在 CPU 环境下处理大型数据集。我们可以采用较小的批次大小在 CPU 环境中处理海量数据。

In [10]:

```
for epoch in range(5):     # 训练整个数据集 5 遍
    for step, (batch_x, batch_y) in enumerate(loader):
# 对于每个训练步骤
# 处理训练数据
        print('Epoch: ', epoch, '| Step: ', step, '| batch x: ',
            batch_x.numpy(), '| batch y: ', batch_y.numpy())
```

Epoch: 0 | Step: 0 | batch x: [3. 2. 4. 7. 8.] | batch y: [8. 9. 7. 4. 3.]
Epoch: 0 | Step: 1 | batch x: [10. 6. 5. 9. 1.] | batch y: [1. 5. 6. 2. 10.]
Epoch: 1 | Step: 0 | batch x: [4. 1. 10. 6. 3.] | batch y: [7. 10. 1. 5. 8.]
Epoch: 1 | Step: 1 | batch x: [7. 5. 8. 9. 2.] | batch y: [4. 6. 3. 2. 9.]
Epoch: 2 | Step: 0 | batch x: [6. 1. 2. 5. 9.] | batch y: [5. 10. 9. 6. 2.]
Epoch: 2 | Step: 1 | batch x: [7. 4. 10. 3. 8.] | batch y: [4. 7. 1. 8. 3.]
Epoch: 3 | Step: 0 | batch x: [2. 3. 1. 10. 7.] | batch y: [9. 8. 10. 1. 4.]
Epoch: 3 | Step: 1 | batch x: [9. 6. 8. 4. 5.] | batch y: [2. 5. 3. 7. 6.]
Epoch: 4 | Step: 0 | batch x: [7. 4. 8. 2. 9.] | batch y: [4. 7. 3. 9. 2.]
Epoch: 4 | Step: 1 | batch x: [1. 10. 5. 3. 6.] | batch y: [10. 1. 6. 8. 5.]

将批次大小设置为 8，再重新训练模型。

In [11]:

```
BATCH_SIZE = 8
loader = Data.DataLoader(
    dataset=torch_dataset,     #torch TensorDataset 格式
```

```
    batch_size=BATCH_SIZE,       # 小批次的大小
    shuffle=True,                # 随机打乱以进行训练
    num_workers=2,               # 用于加载数据的子进程数
)
```

In [12]:

```
for epoch in range(5):          # 训练整个数据集 5 遍
    for step, (batch_x, batch_y) in enumerate(loader):
        # 对于每个训练步骤
        # 处理训练数据
        print('Epoch: ', epoch, '| Step: ', step, '| batch x: ',
              batch_x.numpy(), '| batch y: ', batch_y.numpy())
```

```
Epoch:  0 | Step:  0 | batch x:  [7. 2. 5. 8. 1. 4. 6. 3.] | batch y:
[ 4.  9.  6.  3. 10.  7.  5.  8.]
Epoch:  0 | Step:  1 | batch x:  [10.  9.] | batch y:  [1. 2.]
Epoch:  1 | Step:  0 | batch x:  [ 5.  1.  7.  8. 10.  9.  6.  3.] | batch
y:  [ 6. 10.  4.  3.  1.  2.  5.  8.]
Epoch:  1 | Step:  1 | batch x:  [2. 4.] | batch y:  [9. 7.]
Epoch:  2 | Step:  0 | batch x:  [ 6.  2.  3.  1.  8.  7.  5. 10.] | batch
y:  [ 5.  9.  8. 10.  3.  4.  6.  1.]
Epoch:  2 | Step:  1 | batch x:  [9. 4.] | batch y:  [2. 7.]
Epoch:  3 | Step:  0 | batch x:  [ 4.  3.  5.  7.  2. 10.  6.  1.] | batch
y:  [ 7.  8.  6.  4.  9.  1.  5. 10.]
Epoch:  3 | Step:  1 | batch x:  [8. 9.] | batch y:  [3. 2.]
Epoch:  4 | Step:  0 | batch x:  [ 5.  7.  8. 10.  3.  2.  4.  9.] | batch
y:  [6. 4. 3. 1. 8. 9. 7. 2.]
Epoch:  4 | Step:  1 | batch x:  [6. 1.] | batch y:  [ 5. 10.]
```

秘籍 6-3　确定学习率

问题

如何根据学习率和迭代次数确定最佳解？

解决方案

针对给定的样本张量,采用多种模型进行试验,并输出各模型的参数。学习率和迭代次数的设定与模型准确性密切相关。为了达到损失函数的全局最小值状态,关键在于将学习率保持最小化,迭代次数最大化,从而使迭代过程能够将损失函数降至最低。

编程实战

首先,必须导入相关库。在寻求最小化损失函数的过程中,通常采用梯度下降作为优化算法,这是一个迭代的过程。梯度下降的目标是找到可训练参数的最快下降速率,以实现损失函数的最小化。

In [13]:
```
import torch
import torch.utils.data as Data
import torch.nn.functional as F
from torch.autograd import Variable
import matplotlib.pyplot as plt
%matplotlib inline
```
In [14]:
```
torch.manual_seed(12345)  # 为了可复现
```
In [15]:
```
LR = 0.01
BATCH_SIZE = 32
EPOCH = 12
```

在实验中,所使用的样本数据集涵盖如下范畴(见图6-1)。

In [16]:
```
# 样本数据集
x = torch.unsqueeze(torch.linspace(-1, 1, 1000), dim=1)
y = x.pow(2) + 0.3*torch.normal(torch.zeros(*x.size()))
# 绘制数据集
plt.scatter(x.numpy(), y.numpy())
plt.show()
```

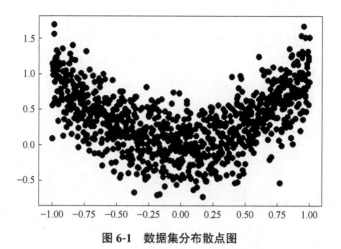

图 6-1　数据集分布散点图

样本数据集的前十条记录如下所示。

In [17] :

x[0:10]

Out [17] :

tensor([[-1.0000], [-0.9980], [-0.9960], [-0.9940], [-0.9920], [-0.9900], [-0.9880], [-0.9860], [-0.9840], [-0.9820]])

In [18] :

y[0:10]

Out [18] :

tensor([[0.5561], [1.1422], [0.0882], [1.1212], [1.0920], [0.9764], [1.0417], [0.5877], [1.6916], [1.5640]])

接下来，可以运用 PyTorch 的 util 函数加载张量数据集，设定批次大小，并对测试数据进行检查。

In [19] :

```
torch_dataset = Data.TensorDataset(x, y)
loader = Data.DataLoader(
    dataset=torch_dataset,
    batch_size=BATCH_SIZE,
    shuffle=True, num_workers=2,)
```

In [20] :

torch_dataset

Out[20]:
<torch.utils.data.dataset.TensorDataset at 0x7f39127b8d50>

In[21]:
```
loader
```
Out[21]:
<torch.utils.data.dataloader.DataLoader at 0x7f39127b8a10>

接下来，定义神经网络模块。

In[22]:
```python
class Net(torch.nn.Module):
    def __init__(self):
        super(Net, self).__init__()
        self.hidden = torch.nn.Linear(1, 20)      # 隐藏层
        self.predict = torch.nn.Linear(20, 1)     # 输出层
    def forward(self, x):
        x = F.relu(self.hidden(x))   # 隐藏层的激活函数
        x = self.predict(x)   # 线性输出
        return x
```

In[23]:
```python
net_SGD         = Net()
net_Momentum    = Net()
net_RMSprop     = Net()
net_Adam        = Net()
nets = [net_SGD, net_Momentum, net_RMSprop, net_Adam]
```

接下来，输出并检查各个网络的模型结构。

In[24]:
```
net_Adam
```
Out[24]:
Net((hidden): Linear(in_features=1, out_features=20, bias=True) (predict): Linear(in_features=20, out_features=1, bias=True))

In[25]:
```
net_Momentum
```
Out[25]:

```
Net( (hidden): Linear(in_features=1, out_features=20, bias=True) (predict):
Linear(in_features=20, out_features=1, bias=True) )
```

In[26]:

```
net_RMSprop
```

Out[26]:

```
Net( (hidden): Linear(in_features=1, out_features=20, bias=True) (predict):
Linear(in_features=20, out_features=1, bias=True) )
```

In[27]:

```
net_SGD
```

Out[27]:

```
Net( (hidden): Linear(in_features=1, out_features=20, bias=True) (predict):
Linear(in_features=20, out_features=1, bias=True) )
```

在优化过程中,可供选择的方案繁多,我们将从中筛选出最佳的方案。

In[28]:

```
opt_SGD         = torch.optim.SGD(net_SGD.parameters(), lr=LR)
opt_Momentum    = torch.optim.SGD(net_Momentum.parameters(),
                                  lr=LR, momentum=0.8)
opt_RMSprop     = torch.optim.RMSprop(net_RMSprop.parameters(),
                                      lr=LR, alpha=0.9)
opt_Adam        = torch.optim.Adam(net_Adam.parameters(),
                                   lr=LR, betas=(0.9, 0.99))
optimizers = [opt_SGD, opt_Momentum, opt_RMSprop, opt_Adam]
```

In[29]:

```
opt_Adam
```

Out[29]:

```
Adam ( Parameter Group 0 amsgrad: False betas: (0.9, 0.99) capturable:
False eps: 1e-08 foreach: None lr: 0.01 maximize: False weight_decay: 0 )
```

In[30]:

```
opt_Momentum
```

Out[30]:

```
SGD ( Parameter Group 0 dampening: 0 foreach: None lr: 0.01 maximize: False
momentum: 0.8 nesterov: False weight_decay: 0 )
```

In[31]:

```
opt_RMSprop
```

Out[31]:

RMSprop (Parameter Group 0 alpha: 0.9 centered: False eps: 1e-08 foreach: None lr: 0.01 momentum: 0 weight_decay: 0)

In[32]:

opt_SGD

Out[32]:

SGD (Parameter Group 0 dampening: 0 foreach: None lr: 0.01 maximize: False momentum: 0 nesterov: False weight_decay: 0)

In[33]:

loss_func = torch.nn.MSELoss()

losses_his = [[], [], [], []]   #记录损失值变化

In[34]:

loss_func

Out[34]:

MSELoss()

在上述代码中，定义了四种不同类型的优化器：SGD（随机梯度下降）、Momentum（带动量的随机梯度下降）、RMSprop（方均根传播）以及Adam（自适应矩估计）。各类优化器均依据特定参数进行初始化，例如学习率（lr）、动量（momentum）、衰减率（alpha）等。

秘籍 6-4　执行并行训练

问题

如何使用PyTorch进行包含大量模型的并行数据训练？

解决方案

优化器实则是一种增强张量的函数。在寻找最优模型过程中，需对众多模型展开并行训练。学习率、批次大小以及优化算法的选取使得各个模型独具特色，

与众不同。超参数优化是挑选最佳模型的关键环节。

编程实战

首先,务必确保导入适当的库。通过调整三个超参数(学习率、批次大小和优化算法),可以实现并行训练多个模型。而最佳模型则根据测试数据集的准确性来确定。以下代码脚本采用了随机梯度下降算法、动量、RMSprop 和 Adam 作为优化方法。

In[35]:

```
# 训练
for epoch in range(EPOCH):
    print('Epoch: ', epoch)
    for step, (batch_x, batch_y) in enumerate(loader):    # 对于每个训练步骤
    b_x = Variable(batch_x)
    b_y = Variable(batch_y)

    for net, opt, l_his in zip(nets, optimizers, losses_his):
    output = net(b_x)              # 获取每个网络的输出
    loss = loss_func(output, b_y)  # 计算每个网络的损失
    opt.zero_grad()                # 清除此训练步骤的梯度
    loss.backward()                # 反向传播,计算梯度
    opt.step()                     # 通过梯度下降执行一步参数更新
    l_his.append(loss.data)        # 记录损失值
labels = ['SGD', 'Momentum', 'RMSprop', 'Adam']
for i, l_his in enumerate(losses_his):
    plt.plot(l_his, label=labels[i])
plt.legend(loc='best')
plt.xlabel('Steps')
plt.ylabel('Loss')
plt.ylim((0, 0.2))
plt.show()
```

接下来看一下图表和迭代情况。

Epoch: 0
Epoch: 1

Epoch: 2
Epoch: 3
Epoch: 4
Epoch: 5
Epoch: 6
Epoch: 7
Epoch: 8
Epoch: 9
Epoch: 10
Epoch: 11

在四种优化器中，RMSprop 优化方法在训练过程中展现出较高的准确性或较小的损失值。如图 6-2 所示。

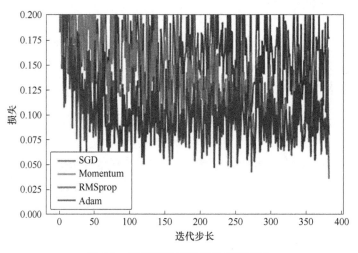

图 6-2　选择不同优化器的训练过程

小结

在本章中，我们深入讨论了使用深度学习模型从训练数据集中进行参数学习的各种策略。通过调整超参数，可以使训练过程更加有效。选用恰当的超参数是一个关键环节。虽然深度学习模型（如卷积神经网络、循环神经网络和深度神经网络）在架构上存在差异，但训练过程以及超参数的选择具有相似性。在 PyTorch 框架中，超参数的选择与优化相较于其他平台更为简便。

第 7 章

使用 PyTorch 进行自然语言处理

自然语言处理（Natural Language Processing，NLP）是指计算机通过对人类语言进行各类任务的研究与评估，以实现对自然语言的理解与生成。这一领域的研究也被称为计算语言学。自然语言处理主要包括两个方面：自然语言理解与自然语言生成。自然语言理解涉及对输入语言的分析和理解，进而作出相应的响应；而自然语言生成则是从输入文本中构建出语言的过程。语言运用方式繁多，一词多义的现象较为常见，因此，消除歧义是自然语言理解的重要环节。

歧义级别可以分为三种类型：

- **词汇歧义**基于词性，即判断一个词是名词、动词、副词等。一个词在不同上下文中可能有不同的含义，例如英语中的"bank"既可以表示银行，也可以表示河岸。
- **句法歧义**是指一个句子可以有多个解释；主语和谓语是中性的，即它们本身并不能提供足够的信息来明确句子的意义。这意味着无法仅依靠主语和谓语的词性或形式来确定句子的具体含义，而需要考虑句子的上下文、语境以及其他语法要素来进行解释和理解。
- **指代歧义**与用词表达的事件或情境有关。在文本中某个词或短语的指代不明确，需要根据上下文来确定其具体指向，例如"他""它"等代词的指代。

文本分析是自然语言处理和理解的前提。文本分析包括语料库的创建（创建一组收集的文档），然后去除空格、标点符号、停用词和无文本含义的垃圾值，如符号、表情符号等。清理完成后，下一步是将文本以向量形式表示。这可以使用标准的 Word2vec 模型完成，或者可以用词频和逆文档频率格式（tf-idf）表示。在当今的世界中，可以看到许多应用程序使用自然语言处理，以下是一些例子：

- 拼写检查应用程序：提供在线和移动设备的拼写检查服务，用户输入单词后，系统会查询单词的含义并建议是否需要纠正拼写。
- 关键字搜索：在过去的十年中，关键字搜索已经成为我们生活中不可或缺的一部分。无论是去餐厅、购物还是出行，我们都会进行在线搜索。如果用户输入的关键字不正确，则不会检索到匹配项。然而如果搜索引擎系统非常智能，便可以预测用户的意图并建议用户实际想要搜索的页面。
- 预测文本：广泛应用于各种聊天应用程序，用户输入一个单词后，系统会根据用户的写作模式，提供一个单词列表供用户选择。用户可以选择列表中的单词来构建句子。
- 问答系统：如 Google Home 和 Amazon Alexa，允许用户使用自然语言与系统交互。系统处理这些信息，进行智能搜索，并检索最佳结果供用户使用。
- 替代数据提取：当用户无法获得实际数据时，可以使用互联网获取公开可用的数据，并搜索相关信息。例如，如果我想买笔记本电脑，想比较不同在线门户的价格。有个系统可以从各个网站抓取价格信息，并提供价格摘要。这个过程被称为替代数据收集，它使用网络抓取、文本处理和自然语言处理等技术。
- 情感分析：对客户、用户或代理表达的文本情感进行分析的过程。可应用于客户评论、电影评论等。需要分析文本并标记为正面情绪或负面情绪。可以通过情感分析构建类似的应用程序。
- 主题建模：从语料库中找出独特主题的过程。例如，从科学、数学、英语和生物学中提取文本并将它们混合在一起，然后要求机器对文本进行分类并告诉我们语料库中有多少个主题等。这也被称为完美的主题建模系统。
- 文本摘要：将语料库中的文本以较短的格式进行总结的过程。如果我们有两页长的文档，包含 1000 个单词，并且需要将其总结成 200 个单词的段落，则可以通过文本摘要算法来实现。
- 语言翻译：将一种语言翻译成另一种语言的过程，例如英语翻译成法语或法语翻译成德语等。语言翻译可以帮助用户理解另一种语言并使沟通过程更加有效。

对人类语言的研究具有离散性，同时也极其复杂。相同的句子可能包含多种不同的含义，但其构建方式却是为了特定的预期受众。为了深入理解自然语言的复杂性，不仅需要借助于工具和程序，还需要掌握系统和方法。在自然语言处理中，以下五个步骤被广泛用于理解用户的文本：

- 词法分析：这一步涉及识别单词的结构。
- 句法分析：是对语法和句法结构的研究。
- 语义分析：探讨一个词在上下文中的字面含义。
- 词性（PoS）分析：理解和解析词性的过程。
- 语用分析：理解一个词在上下文中的实际含义。

在本章中，将使用 PyTorch 实现自然语言处理任务中最常用的步骤。

秘籍 7-1　词嵌入

问题

如何使用 PyTorch 创建词嵌入模型？

解决方案

词嵌入是一种将词汇表中的单词、短语或令牌转换为具有语义意义的向量结构的过程。该过程通过将输入文本映射至实数向量空间，以便后续运用机器学习或深度学习模型进行计算。

编程实战

采用实数向量来表示单词和短语，可以实现更有效的含义相似度计算。在段落或文档中，具有相似含义的单词或短语将被映射至相近的向量表示，从而有助于快速寻找相似单词。当前业界存在众多算法可用于生成文本的嵌入向量，其中最著名的框架包括 Word2vec 和 GloVe。以下为一个具体实例。

In[1]:
```
import torch
```

```python
import torch.nn as nn
import torch.nn.functional as F
import torch.optim as optim

torch.manual_seed(1234)
```

In[2]:
```python
word_to_ix = {"data": 0, "science": 1}
```

In[3]:
```python
word_to_ix
```
Out[3]:
```
{'data': 0, 'science': 1}
```

In[4]:
```python
embeds = nn.Embedding(2, 5)  # 2表示有2个词，5表示5维度
```

In[5]:
```python
embeds
```
Out[5]:
```
Embedding(2, 5)
```

In[6]:
```python
lookup_tensor = torch.tensor([word_to_ix["data"]], dtype=torch.long)
lookup_tensor
```
Out[6]:
```
tensor([0])
```

接下来设置一个嵌入层。

In[7]:
```python
hello_embed = embeds(lookup_tensor)
print(hello_embed)
tensor([[ 0.0461,  0.4024, -1.0115,  0.2167, -0.6123]],
       grad_fn=<EmbeddingBackward0>)
```

In[8]:
```python
CONTEXT_SIZE = 2
```

In[9]:
```python
EMBEDDING_DIM = 10
```

以下是一个示例文本。该文本包含两个段落，每个段落包含多个句子。当我

们将词嵌入技术应用于这两个段落时,可以获得实数向量表示,这些向量可作为后续计算的特征。

In[10]:

test_sentence = """"The popularity of the term "data science" has exploded in business environments and academia, as indicated by a jump in job openings.[32] However, many critical academics and journalists see no distinction between data science and statistics. Writing in Forbes, Gil Press argues that data science is a buzzword without a clear definition and has simply replaced "business analytics" in contexts such as graduate degree programs.[7] In the question-and-answer section of his keynote address at the Joint Statistical Meetings of American Statistical Association, noted applied statistician Nate Silver said, "I think data-scientist is a sexed up term for a statistician....Statistics is a branch of science. Data scientist is slightly redundant in some way and people shouldn't berate the term statistician."[9] Similarly, in business sector, multiple researchers and analysts state that data scientists alone are far from being sufficient in granting companies a real competitive advantage[33] and consider data scientists as only one of the four greater job families companies require to leverage big data effectively, namely: data analysts, data scientists, big data developers and big data engineers.[34]

On the other hand, responses to criticism are as numerous. In a 2014 Wall Street Journal article, Irving Wladawsky-Berger compares the data science enthusiasm with the dawn of computer science. He argues data science, like any other interdisciplinary field, employs methodologies and practices from across the academia and industry, but then it will morph them into a new discipline. He brings to attention the sharp criticisms computer science, now a well respected academic discipline, had to once face.[35] Likewise, NYU Stern's Vasant Dhar, as do many other academic proponents of data science,[35] argues
more specifically in December 2013 that data science is different from the existing practice of data analysis across all disciplines, which focuses only on explaining data sets. Data science seeks actionable and consistent pattern for predictive uses.[1] This practical engineering goal takes data science beyond traditional analytics. Now the data in those disciplines and applied fields that lacked solid theories, like health science and social science, could be sought and utilized to generate powerful predictive

models.[1]""".split()
# 我们应该对输入进行分词,但现在先忽略这一点,构建一个元组列表。
# 每个元组是([word_i-2, word_i-1],目标词)
trigrams = [([test_sentence[i], test_sentence[i + 1]], test_sentence[i + 2])
            for i in range(len(test_sentence)-2)]
# 仅打印前3个,以便查看元组的形式
print(trigrams[:3])

vocab = set(test_sentence)
word_to_ix = {word: i for i, word in enumerate(vocab)}

[(['The', 'popularity'], 'of'), (['popularity', 'of'], 'the'), (['of', 'the'], 'term')]

分词是指将语句分解为若干较小的标记单元的过程,这些单元被称为n-gram。若拆分后的结果为单个单词,则称之为单字;若是两个单词的组合,则称之为双字;若是三个单词的序列,则称之为三字,以此类推。

在 PyTorch 中,n-gram 语言模型具备提取相关关键词的能力。

In [11]:
```
class NGramLanguageModeler(nn.Module):

    def __init__(self, vocab_size, embedding_dim, context_size):
        super(NGramLanguageModeler, self).__init__()
        self.embeddings = nn.Embedding(vocab_size, embedding_dim)
        self.linear1 = nn.Linear(context_size * embedding_dim, 128)
        self.linear2 = nn.Linear(128, vocab_size)

    def forward(self, inputs):
        embeds = self.embeddings(inputs).view((1, -1))
        out = F.relu(self.linear1(embeds))
        out = self.linear2(out)
        log_probs = F.log_softmax(out, dim=1)
        return log_probs
losses = []
loss_function = nn.NLLLoss()
model = NGramLanguageModeler(len(vocab), EMBEDDING_DIM, CONTEXT_SIZE)
optimizer = optim.SGD(model.parameters(), lr=0.001)
```

在上述代码中，n-gram 提取器包含三个参数：词汇表长度（vocab_size）、嵌入向量维度（embedding_dim）以及上下文大小（context_size）。接下来，将设置损失函数并构建模型。

In [12]:

```
model
```

Out [12]:

NGramLanguageModeler((embeddings): Embedding(228, 10) (linear1): Linear(in_features=20, out_features=128, bias=True) (linear2): Linear(in_features=128, out_features=228, bias=True))

接下来，应用 Adam 优化器进行优化操作。

In [13]:

```
optimizer
```

Out [13]:

SGD (Parameter Group 0 dampening: 0 foreach: None lr: 0.001 maximize: False momentum: 0 nesterov: False weight_decay: 0)

在处理文本数据时，提取上下文信息具有重要意义。以下代码可以帮助我们完成这个任务。

In [14]:

```
for epoch in range(10):
    total_loss = 0
    for context, target in trigrams:
        # 步骤 1: 准备模型的输入(即将单词转换为整数索引,并将它们包装在张量中)
        context_idxs = torch.tensor([word_to_ix[w] for w in context], dtype=torch.long)
        # 步骤 2: 由于 torch 会累积梯度,因此在传入新实例前,需要将旧实例的梯度清零
        model.zero_grad()
        # 步骤 3: 执行前向传播,获取下一个单词的对数概率
        log_probs = model(context_idxs)
        # 步骤 4: 计算损失函数(再次提醒,PyTorch 需要目标词包装在一个张量中)
        loss = loss_function(log_probs, torch.tensor([word_to_ix[target]], dtype=torch.long))
        # 步骤 5: 进行反向传播并更新梯度
```

```
        loss.backward()
        optimizer.step()
        # 通过调用 tensor.item() 从单元素的张量中获取 Python 数字
        total_loss += loss.item()
    losses.append(total_loss)
print(losses)  # 训练数据每次迭代损失都减少
```

[1873.4337797164917, 1859.2190294265747, 1845.3114666938782, 1831.6828165054321, 1818.3093104362488, 1805.180431842804, 1792.2873740196228, 1779.6297824382782, 1767.2129256725311, 1755.0498096942902]

上述代码中，在每个训练轮次中，通过迭代训练数据集中的每个上下文和目标词组合，执行以下操作：

● 将上下文中的两个词转换为它们在词典中的索引，并封装为一个长整型的张量，作为模型的输入。

● 调用模型的 zero_grad 方法，清空之前的梯度值，防止梯度累积。

● 调用模型的 forward 方法，传入输入张量，得到输出张量，表示每个词的对数概率分布。

● 调用损失函数，传入输出张量和目标词的索引张量，得到损失值，表示预测结果和真实结果的差异。

● 调用损失值的 backward 方法，计算梯度值，反向传播误差。

● 调用优化器的 step 方法，更新模型的参数，沿着梯度的反方向进行优化。

通过多个轮次的训练，损失值逐渐减小，表示模型在训练数据上的性能提升。

秘籍 7-2　使用 PyTorch 创建 CBOW 模型

问题

如何使用 PyTorch 创建 CBOW 模型？

解决方案

单词和短语可以借助两种方式呈现为向量，即连续词袋（Continuous Bag of

Words，CBOW）和跳字模型（skip gram）。在词袋模型中，通过预测上下文内的单词或短语来学习嵌入向量。所谓上下文，指的是当前单词前后的单词。以一个4字大小的上下文为例，模型会将与当前单词相邻的左侧四个单词和右侧四个单词视为上下文。模型的目标是通过从另一个句子中寻找这八个单词，以预测当前位置的单词。

编程实战

接下来看以下实例。

In [15]:
```
CONTEXT_SIZE = 2   # 左边2个词，右边2个词
```

In [16]:
```
raw_text = """For the future of data science, Donoho projects an
ever-growing environment for open science where data sets used for academic
publications are accessible to all researchers.[36] US National Institute of
Health has already announced plans to enhance reproducibility and
transparency of research data.[39] Other big journals are likewise following
suit.[40][41] This way, the future of data science not only exceeds the
boundary of statistical theories in scale and methodology, but data science
will revolutionize current academia and research paradigms.[36] As Donoho
concludes, "the scope and impact of data science will continue to expand
enormously in coming decades as scientific data and data about science
itself become ubiquitously available."[36]""".split()
```

In [17]:
```
# 通过从'raw_text'中派生出一个集合，我们对数组进行了去重处理
vocab = set(raw_text)
vocab_size = len(vocab)

word_to_ix = {word: i for i, word in enumerate(vocab)}
data = []
for i in range(2, len(raw_text) - 2):
    context = [raw_text[i - 2], raw_text[i - 1],
               raw_text[i + 1], raw_text[i + 2]]
    target = raw_text[i]
    data.append((context, target))
```

```
print(data[:5])
[(['For', 'the', 'of', 'data'], 'future'), (['the', 'future', 'data', 'science', ','], 'of'),
(['future', 'of', 'science', ',', 'Donoho'], 'data'), (['of', 'data', 'Donoho', 'projects'],
'science,'), (['data', 'science,', 'projects', 'an'], 'Donoho')]
```

In [18]:
```
class CBOW(nn.Module):

    def __init__(self):
        pass

    def forward(self, inputs):
        pass

# 创建模型并进行训练。这里的一些函数可以有助于准备数据，以供模块使用
def make_context_vector(context, word_to_ix):
    idxs = [word_to_ix[w] for w in context]
    return torch.tensor(idxs, dtype=torch.long)
make_context_vector(data[0][0], word_to_ix)  # 示例
```

Out[18]:
```
tensor([26, 54, 63, 18])
```

该代码的目标是生成用于连续词袋（CBOW）模型训练的数据集。CBOW模型是一种预测给定上下文中目标词的预测模型，它是Word2Vec模型的一个组成部分。

- 数据预处理：将CONTEXT_SIZE设置为2，以表示上下文的大小，即左侧和右侧各有2个单词。原始文本通过.split()方法被转换为单词列表。

- 构建词汇表：通过将raw_text转换为vocab集合，实现去重操作。vocab_size存储词汇表的大小。word_to_ix是一个字典，将每个单词映射到其在词汇表中的索引。

- 创建训练数据集：通过迭代raw_text，从中提取每个单词的上下文和目标单词，并将它们作为元组（context, target）添加到data列表中。上下文由当前单词左边的两个单词和右边的两个单词组成。

- 定义CBOW模型：创建一个名为CBOW的类，继承自nn.Module，用于定义CBOW模型。在CBOW类中，__init__方法用于初始化模型参数。forward方法是模型的前向传播逻辑，目前还没有具体实现。

- 辅助函数：make_context_vector 函数将给定的上下文转换为索引张量。它接受上下文和 word_to_ix 字典作为输入，并返回一个 Long 型的索引张量。
- 示例：使用第一个样本的上下文调用 make_context_vector 函数，并将结果打印出来。这个例子展示了如何将上下文转换为索引张量。

词袋模型的图形化示意如图 7-1 所示。该模型由三层构成，包括输入层、输出层和投影层。输入层采用词嵌入向量，这一向量兼顾了单词与短语的信息。输出层则对应模型预测的相关单词。投影层是神经网络模型提供的一个计算层。

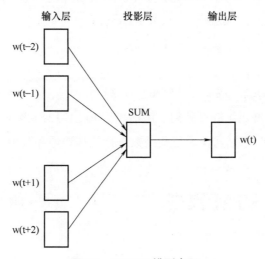

图 7-1 CBOW 模型表示

In [19]:
```
lin = nn.Linear(5, 3)   # 从 R^5 到 R^3 的映射，参数为 A, b
# 数据维度是 2x5，A 将 5 映射到 3。那么我们可以使用 A 对"数据"进行映射吗
data = torch.randn(2, 5)
print(lin(data))  # 事实证明可以进行映射
tensor([[-0.7850,  0.8883,  1.1011], [ 0.3344, -0.3598,  0.5535]],
grad_fn=<AddmmBackward0>)
```

In [20]:
```
data = torch.randn(2, 2)
print(data)
print(F.relu(data))
```

tensor ([[1.6053, -0.1710] , [1.4815, -1.1123]])
tensor ([[1.6053, 0.0000] , [1.4815, 0.0000]])

In [21]:

```
# Softmax 也在 torch.nn.functional 中
data = torch.randn(5)
print(data)
print(F.softmax(data, dim=0))
print(F.softmax(data, dim=0).sum())   # 这是一个分布，因此总和为 1
print(F.log_softmax(data, dim=0))     # functional 中还包含 log_softmax
```

tensor ([-0.4417, -2.5164, -0.2034, -2.1575, -1.2533])
tensor ([0.3313, 0.0416, 0.4204, 0.0596, 0.1471])
tensor (1.0000)
tensor ([-1.1048, -3.1795, -0.8665, -2.8206, -1.9164])

上述代码主要展示了线性映射、激活函数 ReLU 和 Softmax 函数的基本用法，并演示了在 PyTorch 中如何使用这些函数进行数据处理和转换。

秘籍 7-3　LSTM 模型

问题

如何使用 PyTorch 创建 LSTM 模型？

解决方案

长短期记忆（Long Short-Term Memory，LSTM）模型是一种特殊类型的循环神经网络模型，其在自然语言处理领域具有广泛的应用。鉴于文本与句子以序列形式呈现，为构建具有意义的句子，需要采用一种能够记忆文本的长短序列以预测单词或文本的模型。此类模型具备独特的记忆机制，能够有效地处理序列数据。

编程实战

让我们看下面的例子。

In [22]:
```
lstm = nn.LSTM(3, 3)   # 输入维度为3, 输出维度为3
inputs = [torch.randn(1, 3) for _ in range(5)]   # 创建一个长度为5的序列
# 初始化隐藏状态
hidden = (torch.randn(1, 1, 3),
          torch.randn(1, 1, 3))
for i in inputs:
# 逐个遍历序列中的元素
# 每执行一步后,hidden 变量中包含隐藏状态
    out, hidden = lstm(i.view(1, 1, -1), hidden)
inputs = torch.cat(inputs).view(len(inputs), 1, -1)
hidden = (torch.randn(1, 1, 3), torch.randn(1, 1, 3))   # 清除隐藏层状态
out, hidden = lstm(inputs, hidden)
print(out)
print(hidden)
tensor([[[-0.1500,  0.0547,  0.3930]],

        [[-0.1313, -0.0478,  0.0857]],

        [[-0.1131,  0.0047, -0.1003]],

        [[ 0.0176, -0.2464, -0.1589]],

        [[-0.0523,  0.1781, -0.1713]]], grad_fn=<StackBackward0>)
(tensor([[[-0.0523,  0.1781, -0.1713]]], grad_fn=<StackBackward0>),
tensor([[[-0.1997,  0.5137, -0.6064]]], grad_fn=<StackBackward0>))
```

上述代码详尽地展示了如何运用 PyTorch 搭建并运行 LSTM 模型的全过程。LSTM 模型是一种广泛应用于处理序列数据的神经网络模型,尤其在自然语言处理领域发挥出色。以下是具体操作步骤:

- **创建 LSTM 模型**：首先，通过调用 nn.LSTM（input_dim，output_dim）函数创建一个 LSTM 模型，其中 input_dim 代表输入维度，output_dim 代表输出维度。
- **准备输入序列**：接着，创建一个长度为 5 的输入序列 inputs 变量，每个元素是大小为 1×3 的随机张量。
- **初始化隐藏状态**：随后，初始化一个隐藏状态 hidden 变量，包括两个张量，大小为 $1 \times 1 \times 3$，用于存储 LSTM 的隐藏状态。
- **逐步处理序列**：通过 for 循环逐个遍历输入序列 inputs 的每个元素。对于每个元素 i，使用 lstm（i.view（1，1，-1），hidden）进行 LSTM 模型的前向传播，得到输出 out 和更新后的隐藏状态 hidden。
- **整理输入序列和隐藏状态**：在遍历结束后，使用 torch.cat（inputs）.view（len（inputs），1，-1）将所有输入序列拼接起来，并调整维度为序列长度 $1 \times 1 \times 3$。重新初始化隐藏状态 hidden，以清除之前的状态。
- **运行 LSTM 模型**：最后，使用 lstm（inputs，hidden）对整理后的输入序列和隐藏状态进行 LSTM 模型的前向传播，得到输出 out 和最终的隐藏状态 hidden。
- **打印结果**：打印输出结果 out，是一个大小为序列长度 $1 \times 1 \times 3$ 的张量。打印最终的隐藏状态 hidden，包括两个大小为 $1 \times 1 \times 3$ 的张量。

通过以上步骤，展示了如何使用 PyTorch 创建并执行 LSTM 模型，以处理序列数据并获得相应的预测结果和隐藏状态。

接下来，我们准备一个单词序列组，作为 LSTM 网络的训练数据集。

In[23]:
```python
def prepare_sequence(seq, to_ix):
    idxs = [to_ix[w] for w in seq]
    return torch.tensor(idxs, dtype=torch.long)

training_data = [
    ("Probability and random variable are integral part of computation
    ".split(),
     ["DET", "NN", "V", "DET", "NN"]),
    ("Understanding of the probability and associated concepts are
    essential".split(),
     ["NN", "V", "DET", "NN"])
]
```

In[24]:
training_datat

Out[24]:
[(['Probability', 'and', 'random', 'variable', 'are', 'integral', 'part', 'of', 'computation'], ['DET', 'NN', 'V', 'DET', 'NN']), (['Understanding', 'of', 'the', 'probability', 'and', 'associated', 'concepts', 'are', 'essential'], ['NN', 'V', 'DET', 'NN'])]

In[25]:
```
word_to_ix = {}
for sent, tags in training_data:
    for word in sent:
        if word not in word_to_ix:
            word_to_ix[word] = len(word_to_ix)
print(word_to_ix)
tag_to_ix = {"DET": 0, "NN": 1, "V": 2}

EMBEDDING_DIM = 6
HIDDEN_DIM = 6
```

{'Probability': 0, 'and': 1, 'random': 2, 'variable': 3, 'are': 4, 'integral': 5, 'part': 6, 'of': 7, 'computation': 8, 'Understanding': 9, 'the': 10, 'probability': 11, 'associated': 12, 'concepts': 13, 'essential': 14}

In[26]:
```
class LSTMTagger(nn.Module):

    def __init__(self, embedding_dim, hidden_dim, vocab_size, tagset_size):
        super(LSTMTagger, self).__init__()
        self.hidden_dim = hidden_dim

        self.word_embeddings = nn.Embedding(vocab_size, embedding_dim)
# LSTM 将词嵌入作为输入，并输出维度为 hidden_dim 的隐藏状态
        self.lstm = nn.LSTM(embedding_dim, hidden_dim)
        # 将隐藏状态空间映射到标记空间的线性层
        self.hidden2tag = nn.Linear(hidden_dim, tagset_size)
        self.hidden = self.init_hidden()

    def init_hidden(self):
        # 在程序开始之前，没有任何隐藏状态
```

```python
        # 可以参考Pytorch文档，查看为什么具有这种维度
        # 坐标轴语义的维度是(num_layers, minibatch_size, hidden_dim)
        return (torch.zeros(1, 1, self.hidden_dim),
                torch.zeros(1, 1, self.hidden_dim))
    def forward(self, sentence):
        embeds = self.word_embeddings(sentence)
        lstm_out, self.hidden = self.lstm(
            embeds.view(len(sentence), 1, -1), self.hidden)
        tag_space = self.hidden2tag(lstm_out.view(len(sentence), -1))
        tag_scores = F.log_softmax(tag_space, dim=1)
        return tag_scores
```

In [27]:

```python
model = LSTMTagger(EMBEDDING_DIM, HIDDEN_DIM, len(word_to_ix), len(tag_to_ix))
loss_function = nn.NLLLoss()
optimizer = optim.SGD(model.parameters(), lr=0.1)
model
loss_function
optimizer
```

Out[27]:

SGD (Parameter Group 0 dampening: 0 foreach: None lr: 0.1 maximize: False momentum: 0 nesterov: False weight_decay: 0)

In [28]:

```python
with torch.no_grad():
    inputs = prepare_sequence(training_data[0][0], word_to_ix)
    tag_scores = model(inputs)
    print(tag_scores)
```

```
tensor([[-1.0414, -1.1928, -1.0680],
        [-1.0747, -1.2163, -1.0154],
        [-1.0706, -1.2298, -1.0083],
        [-1.0661, -1.2428, -1.0022],
        [-1.0013, -1.2948, -1.0254],
        [-1.0539, -1.2640, -0.9973],
        [-1.0718, -1.2705, -0.9757],
        [-0.9919, -1.2527, -1.0689],
        [-0.9726, -1.2880, -1.0611]])
```

在上述代码段中，我们构建了一个基础的序列标注模型，其作用是将输入的句子序列映射为相应的标签序列。该模型以深度学习框架为基础，并采纳了循环神经网络中的 LSTM 结构，实现了以下功能：

- 首先，在数据准备阶段，设置训练数据集 training_data，包括输入的句子和对应的标记序列。同时，为了方便模型训练过程中的矩阵运算，还构建了单词到索引的映射。

- 接下来，在模型定义阶段，采用了 LSTM 模型进行序列标记任务。该模型包括词嵌入层（self.word_embeddings）、LSTM 层（self.lstm）和线性层（self.hidden2tag）三个主要组成部分。词嵌入层将输入的单词转换为高维空间的向量表示，LSTM 层通过模拟人类记忆机制对序列数据进行建模，线性层则用于将 LSTM 层的输出映射到标记空间。最后，对标记空间的得分应用 log softmax 函数，得到每个时间戳的标记得分（概率）。

- 在前向传播和训练阶段，通过执行前向传播算法，将输入的句子序列转换为预测的标记序列。为了衡量模型的预测性能，采用负对数似然损失函数（NLLLoss）来计算模型的损失值。随后，通过梯度下降优化器更新模型的参数，以便在训练数据上进行迭代训练。

- 最后，在预测结果展示阶段，我们打印了预测的标记得分，展示了模型对输入句子序列的预测结果。这些预测结果可以帮助我们评估模型的性能和优化模型的参数配置。

通过以上步骤，我们实现了一个序列标记模型。

小结

本章详尽地讨论了如何运用连续词袋、词嵌入以及构建长短期记忆网络的方法。各节均附有相应的 PyTorch 代码，旨在便于搭建高效且可靠的自然语言处理流水线，进而应用于文本分类、自动文本摘要及情感分析等诸多领域。下一章将介绍分布式 PyTorch，并探讨如何运用其实现大规模并行处理，以及优化 PyTorch 函数和例程。此外，还将学习深度学习模型量化策略，以减小模型体积并提高模型在实际部署环境中的性能表现。

第 8 章

分布式 PyTorch 建模、模型优化和部署

本章将深入探讨如何利用 PyTorch 实现分布式模型训练中的常见安装、训练和设置流程。对于分布式**数据并行训练**和**模型并行训练**的架构，可以参见本章中的图示来理解。分布式模型优化的目标是缩减模型参数的规模，从而使模型更为轻便。然而，模型规模越大，推理生成过程则相对缓慢。如果减少深度学习模型中的层数，那么模型训练的参数也会减少，但这可能会影响到模型的准确度。因此，为缩小模型尺寸，业界提出了一种名为量化的技术。在将模型投入实际应用前，需采用各类模型量化策略。否则，较大的模型将难以适应实际部署需求。

秘籍 8-1 分布式 Torch 架构

问题

什么是分布式 Torch 架构，它们是如何设计的？

解决方案

深度学习模型的训练负载可以分布在多个 GPU 和 CPU 上。分布式训练方法主要有两种：一种方式是将训练数据分布在多个处理器上进行处理，即**数据并行**；另一种方式是将梯度分布在多个处理器上，即**模型并行**。

图 8-1 演示了训练数据样本如何通过划分成较小的批次数据（如 Mini Batch1 和 Mini Batch2）进行分布式训练。部分数据子样本被分配至集群中的机器，这些机器集群由多个处理器组合而成，以实现分布式模型训练。在 Machine1 中，深度学习模型设有四个隐藏层。当 Mini Batch1 的数据经过这四层后，估算损失函数值。同样地，在将 Mini Batch2 输入 Machine2 时，也遵循相同流程并再次估算损失函数值。根据损失值，梯度更新传递到两台机器的各层，并使用反向传播算法更新梯度以优化模型。

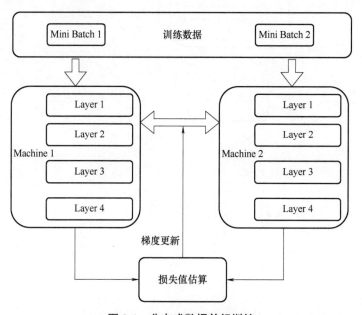

图 8-1 分布式数据并行训练

在图 8-2 中，训练数据被划分为若干较小的数据批次。这些小批次数据经过 Machine1 的四个隐藏层处理后，传入 Machine2 的相同数量的隐藏层，接着进行损失值估算和梯度更新，并行发送至两台机器。此过程使得梯度处理速度相较于图 8-1 所示方法要快一些。

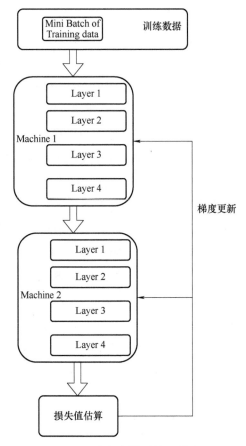

图 8-2　分布式模型并行训练

秘籍 8-2　Torch 分布式组件

问题

Torch 分布式的组件是什么？

解决方案

Torch distributed 是一个用于支持多进程并行计算的 PyTorch 包，可以在一个或多个机器上运行。该库的功能涵盖了三个主要组件：

- **分布式数据并行训练**是一种广泛应用的单程序多数据训练范式，通过在每个进程上复制模型，并输入不同的数据样本，同时负责梯度通信以保持模型副本的同步，并与梯度计算重叠以加速训练。
- **基于远程过程调用的分布式训练**：基于远程过程调用的分布式训练适用于无法适配 DDP 框架的工作负载。一个例子是并行分布式流水线处理。
- **集体通信**：存在一个名为 c10d 的库，提供集合通信和点对点通信功能。

有三种可以使用的后端：GLOO、NCCL 和 MPI。对于分布式 GPU 训练，应使用 NCCL 后端；对于分布式 CPU 训练，应使用 GLOO 后端；如果 PyTorch 是从源代码构建的，则应使用 MPI 后端。

秘籍 8-3　设置分布式 PyTorch

问题

如何在 PyTorch 中设置分布式并行处理？

解决方案

在 PyTorch 中，分布式训练有两种配置方式：云端 GPU 环境和本地 CPU 环境。数据并行训练需要按照一组步骤进行设置，以实现框架的最佳利用。

编程实战

以下代码示例展示了如何导入所需库以安装 DDP，并设置环境以在本地 55555 端口运行。在此示例中，分布式进程组需配备一个后端，本例选用 GLOO。此外，还需提供用户指定的 rank 和 worker 数量（即 world_size）。

In[1]:
```
import os
import sys
```

```python
import tempfile
import torch
import torch.distributed as dist
import torch.nn as nn
import torch.optim as optim
import torch.multiprocessing as mp
```
In[2]:
```python
from torch.nn.parallel import DistributedDataParallel as DDP
```
In[3]:
```python
def setup(rank, world_size):
    os.environ['MASTER_ADDR'] = 'localhost'
    os.environ['MASTER_PORT'] = '5555'

    # 初始化进程组
    dist.init_process_group("gloo", rank=rank, world_size=world_size)

def cleanup():
    dist.destroy_process_group()
```
In[4]:
```python
class NNET_Model(nn.Module):
    def __init__(self):
        super(NNET_Model, self).__init__()
        self.net1 = nn.Linear(10, 10)
        self.relu = nn.ReLU()
        self.net2 = nn.Linear(10, 5)

    def forward(self, x):
        return self.net2(self.relu(self.net1(x)))
```
In[5]:
```python
def nnet_basic(rank, world_size):
    print(f"Running basic DDP example on rank {rank}.")
    setup(rank, world_size)

    # 创建模型并将其移动到具有id为rank的CPU上
    model = NNET_Model().to(rank)
    ddp_model = DDP(model, device_ids=[rank])

    loss_fn = nn.MSELoss()
    optimizer = optim.SGD(ddp_model.parameters(), lr=0.001)
```

```
optimizer.zero_grad()
outputs = ddp_model(torch.randn(20, 10))
labels = torch.randn(20, 5).to(rank)
loss_fn(outputs, labels).backward()
optimizer.step()

cleanup()
```

在基本的神经网络模型之后,引入一个分布式数据并行(DDP)层。

In[6]:

```
nnet_basic(rank=1,world_size=4)
```

在 rank 为 1 的 GPU 上运行基本的 nnet_basic 实例,其中 world_size 的值为 4,表示一共有四台工作站或四块 GPU 参与训练任务。

秘籍 8-4 加载数据到分布式 PyTorch

问题

如何将数据集加载到分布式 PyTorch 中?

解决方案

本节将展示下载 MNIST 数据集的过程,以及利用分布式数据采样器进行数据准备和模型训练的方法。

编程实战

让我们看下面的代码示例。

In[7]:

```
import torch.distributed as dist
def setup(rank, world_size):
    os.environ['MASTER_ADDR'] = 'localhost'
    os.environ['MASTER_PORT'] = '12355'
```

```
        dist.init_process_group("nccl", rank=rank, world_size=world_size)
```
In[8]:
```
import torchvision.datasets as datasets
```
In[9]:
```
mnist_trainset = datasets.MNIST(root='./data', train=True, download=True,
transform=None)
```

Downloading http://yann.lecun.com/exdb/mnist/train-images-idx3-ubyte.gz
Downloading http://yann.lecun.com/exdb/mnist/train-images-idx3-ubyte.gz to
./data/MNIST/raw/train-images-idx3-ubyte.gz
100%
9912422/9912422 [00:00<00:00, 120503370.11it/s]
Extracting ./data/MNIST/raw/train-images-idx3-ubyte.gz to ./data/MNIST/raw

Downloading http://yann.lecun.com/exdb/mnist/train-labels-idx1-ubyte.gz
Downloading http://yann.lecun.com/exdb/mnist/train-labels-idx1-ubyte.gz to
./data/MNIST/raw/train-labels-idx1-ubyte.gz
100%
28881/28881 [00:00<00:00, 798789.93it/s]
Extracting ./data/MNIST/raw/train-labels-idx1-ubyte.gz to ./data/MNIST/raw

Downloading http://yann.lecun.com/exdb/mnist/t10k-images-idx3-ubyte.gz
Downloading http://yann.lecun.com/exdb/mnist/t10k-images-idx3-ubyte.gz to
./data/MNIST/raw/t10k-images-idx3-ubyte.gz
100%
1648877/1648877 [00:00<00:00, 47304965.85it/s]
Extracting ./data/MNIST/raw/t10k-images-idx3-ubyte.gz to ./data/MNIST/raw

Downloading http://yann.lecun.com/exdb/mnist/t10k-labels-idx1-ubyte.gz
Downloading http://yann.lecun.com/exdb/mnist/t10k-labels-idx1-ubyte.gz to
./data/MNIST/raw/t10k-labels-idx1-ubyte.gz
100%
4542/4542 [00:00<00:00, 183349.17it/s]
Extracting ./data/MNIST/raw/t10k-labels-idx1-ubyte.gz to ./data/MNIST/raw

In[10]:
```
from torch.utils.data.distributed import DistributedSampler
def prepare(rank, world_size, batch_size=32, pin_memory=False,
num_workers=0):
    dataset = mnist_trainset
```

```python
    sampler = DistributedSampler(dataset, num_replicas=world_size,
    rank=rank, shuffle=False, drop_last=False)

    dataloader = DataLoader(dataset, batch_size=batch_size, pin_memory=
    pin_memory, num_workers=num_workers, drop_last=False, shuffle=False,
    sampler=sampler)

    return dataloader
```

In[11]:
```python
from torch.nn.parallel import DistributedDataParallel as DDP
def main(rank, world_size):
    # 设置进程组
    setup(rank, world_size)
    # 准备数据加载器
    dataloader = prepare(rank, world_size)
    # 实例化模型(此处是自己的模型)并将其移动到正确的设备
    model = Model().to(rank)
    # 使用DDP包装模型
    # device_ids 向DDP传递模型在哪个设备上
    # output_device 向DDP传递模型输出到哪个设备,在本例子中为rank
    # find_unused_parameters=True 表示DDP将查找模型中所有forward()函数未输
        出的模块
    model = DDP(model, device_ids=[rank], output_device=rank, find_unused_
    parameters=True)
```

In[12]:
```python
if torch.cuda.is_available():
    DEVICE = torch.device('cuda')
    device_ids = list(range(torch.cuda.device_count()))
    gpus = len(device_ids)
    print('GPU detected')
else:
    DEVICE = torch.device("cpu")
    print('No GPU. switching to CPU')
```

上述代码的核心目标是实现在分布式环境下加载MNIST数据集,并借助分布式数据并行训练模型。具体操作步骤如下:

- 设置分布式进程组：使用 dist.init_process_group( ) 函数设置分布式进程组。这里使用"nccl"作为后端，rank 表示当前进程的排名，world_size 表示总共的进程数。
- 准备数据加载器：使用 torchvision.datasets.MNIST 类加载 MNIST 数据集，并使用 DistributedSampler 类创建分布式采样器，用于在分布式环境中对数据进行采样和分发。然后使用 torch.utils.data.DataLoader 类创建数据加载器，将数据集、批量大小、是否固定内存、工作线程数等参数传入。
- 主函数：在 main( ) 函数中，首先调用 setup( ) 函数设置分布式进程组。然后调用 prepare( ) 函数准备数据加载器。接着创建自定义模型，并将模型移动到适当的设备上。最后使用 torch.nn.parallel.DistributedDataParallel 类对模型进行封装，以实现分布式数据并行训练。
- 检测 GPU 可用性：使用 torch.cuda.is_available( ) 函数检测是否有可用的 GPU 设备。如果有 GPU 设备，则将 DEVICE 设置为 cuda，并获取可用的 GPU 设备 ID 列表。如果没有 GPU 设备，则将 DEVICE 设置为 CPU。

秘籍 8-5　PyTorch 中的模型量化

问题

如何在 PyTorch 中优化深度学习模型？

解决方案

为了在服务器及边缘设备上实现深度学习模型的高效部署，PyTorch 提供了一个名为模型量化的框架。该框架采用 8 位整数格式，旨在降低权重存储需求和计算复杂度，因为原始权重通常采用 32 位或 64 位浮点数格式。通过运用量化策略，PyTorch 模型生成的推理速度可实现显著提升。简单来说，量化是一种利用较低精度数据进行计算和内存访问的技术。

编程实战

在 PyTorch 中，包含以下三种类型的量化方法可供选择：

- **动态量化**是一种神经网络模型优化方法，它要求将模型中每个隐藏层的权重和偏置，以及在进行计算之前的激活函数，这两组数值信息转换为 int8 类型。动态量化在权重和偏置层以及在进入计算层之前的激活函数上都应用了量化技术。由于这种方法是在模型训练过程中应用的，因此被称为动态的。

In [13]:
```
import torch.quantization
quantized_model = torch.quantization.quantize_dynamic(model,
                                                      {torch.nn.Linear},
                                                      dtype=torch.qint8)
```

In [14]:
```
print(quantized_model)
ToyModel(
  (net1): DynamicQuantizedLinear(in_features=10, out_features=10, dtype=torch.qint8, qscheme=torch.per_tensor_affine)
  (relu): ReLU()
  (net2): DynamicQuantizedLinear(in_features=10, out_features=5, dtype=torch.qint8, qscheme=torch.per_tensor_affine)
)
```

- **静态量化技术**是在生成模型对象以及深度学习模型训练完成后实施的。该技术主要由以下三个核心组件构成，这些组件将量化方法应用于模型：
  - 观察器：在训练数据输入模型时，量化过程会在每个激活点上对不同激活函数的结果分布进行检查。这些组件被称为观察器，因为它们在应用量化时会观察和记录不同激活函数的结果分布以及与这些分布相关的统计数据的变化。
  - 融合：运算符融合是将多个操作合并为单个操作的特性，从而减少计算量。
  - 通道预量化：此特性有助于量化输出通道并加快计算速度。

In[15]:
# 插入观察器
torch.quantization.prepare(model, inplace=True)
# 校准模型并收集统计信息
ToyModel((net1): Linear(in_features=10, out_features=10, bias=True) (relu): ReLU() (net2): Linear(in_features=10, out_features=5, bias=True))

In[16]:
# 转换为量化版本
torch.quantization.convert(model, inplace=True)
ToyModel((net1): Linear(in_features=10, out_features=10, bias=True) (relu): ReLU() (net2): Linear(in_features=10, out_features=5, bias=True))

- **量化感知训练**（QAT）：此方法旨在提高模型准确性，其运作原理在于，在前向传播、反向传播以及激活应用函数的过程中，将浮点数值近似截断为int8，但实际上依然保持了浮点数的精度。这种近似处理方式有助于提升模型性能。

In[17]:
# 准备量化感知训练
torch.quantization.prepare_qat(model, inplace=True)

ToyModel((net1): Linear(in_features=10, out_features=10, bias=True) (relu): ReLU() (net2): Linear(in_features=10, out_features=5, bias=True))

In[18]:
# 转换为量化版本，移除dropout，以检查每个模型的准确度
epochquantized_model=torch.quantization.convert(model.eval(), inplace=False)

In[19]:
epochquantized_model
ToyModel((net1): Linear(in_features=10, out_features=10, bias=True) (relu): ReLU() (net2): Linear(in_features=10, out_features=5, bias=True))

秘籍 8-6　量化观察器应用

问题

如何在 PyTorch 中应用不同的量化观察器？

解决方案

在每一层计算之前，我们会使用不同类型的观察器来监测激活分布的情况。因此，必须熟知每种类型观察器的操作特性及工作原理。

编程实战

在 PyTorch 中，有三种不同类型的量化观察器：

- **MinMaxObserver**（最小-最大观察器）：这种方法使用最小值和最大值来计算量化参数，即缩放因子和零点。它适用于对称和非对称量化，以及无符号和有符号整数类型。
- **MovingAverageMinMaxObserver**（移动平均最小-最大观察器）：这种方法与 MinMaxObserver 类似，但是使用移动平均值来跟踪最小值和最大值，从而减少噪声和异常值的影响。
- **HistogramObserver**（直方图观察器）：这种方法使用直方图来估计数据的分布，并根据分布的累积密度函数来计算量化参数。它适用于非对称量化，以及无符号和有符号整数类型。

不同类型的量化观察器可以帮助用户更好地理解和分析深度学习模型中的数据。

In [20]:
```
from torch.quantization.observer import MinMaxObserver,
MovingAverageMinMaxObserver, HistogramObserver
C, L = 5, 5
```

```python
normal = torch.distributions.normal.Normal(0,1)
inputs = [normal.sample((C, L)), normal.sample((C, L))]
print(inputs)
[tensor([[ 0.8052, -0.1585, -1.5735,  0.0400, -0.1424],
        [-1.4450,  1.2916, -0.4354, -1.8434,  0.4686],
        [-1.6375,  0.0545,  0.5203,  0.0024,  0.7699],
        [-1.2877, -2.1810,  0.4022,  1.3470,  0.9177],
        [-0.5629, -0.5823, -1.0329, -1.3076,  0.9457]]),
 tensor([[ 0.3280, -1.9777,  0.2115,  0.8891,  1.2109],
        [-0.0630, -0.4131, -0.3992, -0.4765, -0.7934],
        [ 0.7557, -0.7131, -1.6143, -0.9568,  0.4245],
        [ 0.0509,  0.1589,  0.9872,  1.1071, -0.0961],
        [-0.7442,  1.6635, -0.2982, -0.4168,  0.2499]])]
```

In [21]:

```python
observers = [MinMaxObserver(), MovingAverageMinMaxObserver(), HistogramObserver()]
for obs in observers:
    for x in inputs: obs(x)
    print(obs.__class__.__name__, obs.calculate_qparams())
MinMaxObserver (tensor([0.0151]), tensor([145], dtype=torch.int32))
MovingAverageMinMaxObserver (tensor([0.0138]), tensor([157], dtype=torch.int32))
HistogramObserver (tensor([0.0150]), tensor([143], dtype=torch.int32))
```

In [22]:

```python
from torch.quantization.observer import MovingAveragePerChannelMinMaxObserver
obs = MovingAveragePerChannelMinMaxObserver(ch_axis=0)
# 分别计算所有 'C' 通道的 qparams
for x in inputs: obs(x)
print(obs.calculate_qparams())

(tensor([0.0094, 0.0122, 0.0094, 0.0137, 0.0088]), tensor([169, 150, 173, 157, 147], dtype=torch.int32))
```

秘籍 8-7　使用 MNIST 数据集应用量化技术

问题

如何在 PyTorch 中使用 MNIST 数据集对卷积神经网络模型应用不同的量化技术？

解决方案

首先，我们需要将数据集加载到会话中，以便在后续处理中能够随时使用。接着，需要应用必要的脚本，来启动模型训练的过程。最后，应用量化技术。

编程实战

以下是一段代码示例。

```
In [23]:
import torch
import torchvision
import torchvision.transforms as transforms
import torch.nn as nn
import torch.nn.functional as F
import torch.optim as optim
import os
from torch.utils.data import DataLoader
import torch.quantization
from torch.quantization import QuantStub, DeQuantStub

In [24]:
transform = transforms.Compose(
    [transforms.ToTensor(),
     transforms.Normalize((0.5,), (0.5,))])

trainset = torchvision.datasets.MNIST(root='./data', train=True,
```

```python
                                    download=True, transform=transform)
trainloader = torch.utils.data.DataLoader(trainset, batch_size=64,
                                    shuffle=True, num_workers=16,
                                    pin_memory=True)

testset = torchvision.datasets.MNIST(root='./data', train=False,
                                    download=True, transform=transform)
testloader = torch.utils.data.DataLoader(testset, batch_size=64,
                                    shuffle=False, num_workers=16,
                                    pin_memory=True)
```

下面的 AverageMeter 类计算并存储权重的平均值和当前值。

In [25]:
```python
class AverageMeter(object):
    """Computes and stores the average and current value"""
    def __init__(self, name, fmt=':f'):
        self.name = name
        self.fmt = fmt
        self.reset()

    def reset(self):
        self.val = 0
        self.avg = 0
        self.sum = 0
        self.count = 0

    def update(self, val, n=1):
        self.val = val
        self.sum += val * n
        self.count += n
        self.avg = self.sum / self.count

    def __str__(self):
        fmtstr = '{name} {val' + self.fmt + '} ({avg' + self.fmt + '})'
        return fmtstr.format(**self.__dict__)
```

为了计算基于 MNIST 数据集的卷积神经网络模型的准确率，可以采用如下函数。

In [26]:
```python
def accuracy(output, target):
    """ Computes the top 1 accuracy """
```

```
with torch.no_grad():
    batch_size = target.size(0)

    _, pred = output.topk(1, 1, True, True)
    pred = pred.t()
    correct = pred.eq(target.view(1, -1).expand_as(pred))

    res = []
    correct_one = correct[:1].view(-1).float().sum(0, keepdim=True)
    return correct_one.mul_(100.0 / batch_size).item()
```

以下脚本可用于输出模型对象的大小（以 MB 为单位）。

In [27]：

```
def print_size_of_model(model):
    """ Prints the real size of the model """
    torch.save(model.state_dict(), "temp.p")
    print('Size (MB):', os.path.getsize("temp.p")/1e6)
    os.remove('temp.p')
```

将量化模型和真实模型作为两个对象放在一起。

```
def load_model(quantized_model, model):
    """ Loads in the weights into an object meant for quantization """
    state_dict = model.state_dict()
    model = model.to('cpu')
    quantized_model.load_state_dict(state_dict)

def fuse_modules(model):
    """ Fuse together convolutions/linear layers and ReLU """
    torch.quantization.fuse_modules(model, [['conv1', 'relu1'],
                                            ['conv2', 'relu2'],
                                            ['fc1', 'relu3'],
                                            ['fc2', 'relu4']],
                                            inplace=True)
```

在上述代码中，load_model（ ）函数的主要作用是将预先训练好的模型权重迁移至一个用于量化的模型对象中。而 fuse_modules（ ）函数则是用于将模型中的卷积层或线性层与 ReLU 层整合，从而降低计算负担和内存占用。

在接下来的代码片段中，将构建一个包含两个卷积层和三个全连接层的神经网络模型进行训练。

In [28]:
```python
class Net(nn.Module):
    def __init__(self, q = False):
        # 通过设置变量q, 可以开启/关闭量化
        super(Net, self).__init__()
        self.conv1 = nn.Conv2d(1, 6, 5, bias=False)
        self.relu1 = nn.ReLU()
        self.pool1 = nn.MaxPool2d(2, 2)
        self.conv2 = nn.Conv2d(6, 16, 5, bias=False)
        self.relu2 = nn.ReLU()
        self.pool2 = nn.MaxPool2d(2, 2)
        self.fc1 = nn.Linear(256, 120, bias=False)
        self.relu3 = nn.ReLU()
        self.fc2 = nn.Linear(120, 84, bias=False)
        self.relu4 = nn.ReLU()
        self.fc3 = nn.Linear(84, 10, bias=False)
        self.q = q
        if q:
            self.quant = QuantStub()
            self.dequant = DeQuantStub()
    def forward(self, x: torch.Tensor) -> torch.Tensor:
        if self.q:
            x = self.quant(x)
        x = self.conv1(x)
        x = self.relu1(x)
        x = self.pool1(x)
        x = self.conv2(x)
        x = self.relu2(x)
        x = self.pool2(x)
        # 注意, 此处要使用reshape而不是view
        x = x.reshape(x.shape[0], -1)
        x = self.fc1(x)
        x = self.relu3(x)
        x = self.fc2(x)
        x = self.relu4(x)
        x = self.fc3(x)
        if self.q:
```

```python
        x = self.dequant(x)
        return x
```

经过训练后的模型对象大小为 0.178587MB。

In[29]:
```python
net = Net(q=False)
print_size_of_model(net)
Size (MB): 0.178587
```

在接下来的步骤中，我们将定义训练函数。在进行训练过程中，将采用先前定义的 AverageMeter 类来计算损失值和准确率。

In[30]:
```python
def train(model: nn.Module, dataloader: DataLoader, cuda=False, q=False):
    criterion = nn.CrossEntropyLoss()
    optimizer = optim.SGD(model.parameters(), lr=0.001, momentum=0.9)
    model.train()
    for epoch in range(20):  # 循环遍历数据集多次
        running_loss = AverageMeter('loss')
        acc = AverageMeter('train_acc')
        for i, data in enumerate(dataloader, 0):
            # 获取输入数据；数据是一个包含[输入,标签]的列表
            inputs, labels = data
            if cuda:
              inputs = inputs.cuda()
              labels = labels.cuda()
            # 将参数梯度清零
            optimizer.zero_grad()

            if epoch>=3 and q:
              model.apply(torch.quantization.disable_observer)
            # 前向传播、反向传播、参数优化
            outputs = model(inputs)
            loss = criterion(outputs, labels)
            loss.backward()
            optimizer.step()
            # 打印输出统计信息
            running_loss.update(loss.item(), outputs.shape[0])
```

```
            acc.update(accuracy(outputs, labels), outputs.shape[0])
        if i % 100 == 0:  # 每 100 个小批量打印一次
            print('[%d, %5d] ' %
                  (epoch + 1, i + 1), running_loss, acc)
    print('Finished Training')
```

接下来，我们将定义测试脚本函数。

In [31]:
```
def test(model: nn.Module, dataloader: DataLoader, cuda=False) -> float:
    correct = 0
    total = 0
    model.eval()
    with torch.no_grad():
        for data in dataloader:
            inputs, labels = data
            if cuda:
                inputs = inputs.cuda()
                labels = labels.cuda()
            outputs = model(inputs)
            _, predicted = torch.max(outputs.data, 1)
            total += labels.size(0)
            correct += (predicted == labels).sum().item()

    return 100 * correct / total
```

在接下来的环节中，我们将调用训练脚本以完成神经网络的训练，并在测试脚本的辅助下进行系统测试。

In [32]:
```
train(net, trainloader)
[15,   301] loss 0.044805 (0.041544) train_acc 98.437500 (98.681478)
[15,   401] loss 0.089017 (0.040428) train_acc 98.437500 (98.753117)
[15,   501] loss 0.001939 (0.041203) train_acc 100.000000 (98.727545)
[15,   601] loss 0.031541 (0.042560) train_acc 98.437500 (98.679285)
[15,   701] loss 0.047192 (0.042918) train_acc 96.875000 (98.684914)
[15,   801] loss 0.011530 (0.043959) train_acc 100.000000 (98.642322)
[15,   901] loss 0.030178 (0.044269) train_acc 98.437500 (98.638665)
[16,     1] loss 0.006916 (0.006916) train_acc 100.000000 (100.000000)
```

In[33]:
```
score = test(net, testloader, cuda=False)
print('Accuracy of the network on the test images: {}% - FP32'.
format(score))
Accuracy of the network on the test images: 98.65% - FP32
```

采用浮点权重类型的当前卷积神经网络模型，其精确度为 98.65%。接下来，我们将应用模型量化技术对模型进行优化，并观察其对模型大小和准确率的影响。

In[34]:
```
# 训练后量化
```

In[35]:
```
qnet = Net(q=True)
load_model(qnet, net)
fuse_modules(qnet)
```

In[36]:
```
print_size_of_model(qnet)
score = test(qnet, testloader, cuda=False)
print('Accuracy of the fused network on the test images: {}% - FP32'.
format(score))
Size (MB): 0.178843
Accuracy of the fused network on the test images: 98.65% - FP32
```

应用 fuse 后，模型对象保持 0.178843MB。将训练好的模型 net 参数加载到 qnet 中，然后应用模型融合，模型对象的大小变化不大，且准确率保持不变。

In[37]:
```
qnet.qconfig = torch.quantization.default_qconfig
print(qnet.qconfig)

torch.quantization.prepare(qnet, inplace=True)
print('Post Training Quantization Prepare: Inserting Observers')
print('\n Conv1: After observer insertion \n\n', qnet.conv1)

test(qnet, trainloader, cuda=False)
print('Post Training Quantization: Calibration done')
torch.quantization.convert(qnet, inplace=True)
print('Post Training Quantization: Convert done')
print('\n Conv1: After fusion and quantization \n\n', qnet.conv1)
```

```
print("Size of model after quantization")
print_size_of_model(qnet)
QConfig(activation=functools.partial(<class 'torch.ao.quantization.
observer.MinMaxObserver'>, quant_min=0, quant_max=127){}, weight=functools.
partial(<class 'torch.ao.quantization.observer.MinMaxObserver'>,
dtype=torch.qint8, qscheme=torch.per_tensor_symmetric){})
Post Training Quantization Prepare: Inserting Observers

 Conv1: After observer insertion

 ConvReLU2d(
   (0): Conv2d(1, 6, kernel_size=(5, 5), stride=(1, 1), bias=False)
   (1): ReLU()
   (activation_post_process): MinMaxObserver(min_val=inf, max_val=-inf)
 )
Post Training Quantization: Calibration done
Post Training Quantization: Convert done

 Conv1: After fusion and quantization

 QuantizedConvReLU2d(1, 6, kernel_size=(5, 5), stride=(1, 1),
scale=0.06902680546045303, zero_point=0, bias=False)
Size of model after quantization
Size (MB): 0.049714
```

In [38]:

```
score = test(qnet, testloader, cuda=False)
print('Accuracy of the fused and quantized network on the test images:
{}% - INT8'.format(score))
Accuracy of the fused and quantized network on the test images:
98.58% - INT8
```

经过量化处理，模型尺寸显著缩减至仅 0.05MB，同时准确率并未明显下降。然而，通常情况下，量化处理可能导致模型准确率降低。为了解决这一问题，可以更改默认设置并自定义观察器。

In [39]:

```
from torch.quantization.observer import MovingAverageMinMaxObserver

qnet = Net(q=True)
load_model(qnet, net)
```

```python
fuse_modules(qnet)

qnet.qconfig = torch.quantization.QConfig(
                    activation=MovingAverageMinMaxObserver.with_
                    args(reduce_range=True),
                    weight=MovingAverageMinMaxObserver.with_args
                    (dtype=torch.qint8, qscheme=torch.per_tensor_symmetric))
print(qnet.qconfig)
torch.quantization.prepare(qnet, inplace=True)
print('Post Training Quantization Prepare: Inserting Observers')
print('\n Conv1: After observer insertion \n\n', qnet.conv1)

test(qnet, trainloader, cuda=False)
print('Post Training Quantization: Calibration done')
torch.quantization.convert(qnet, inplace=True)
print('Post Training Quantization: Convert done')
print('\n Conv1: After fusion and quantization \n\n', qnet.conv1)
print("Size of model after quantization")
print_size_of_model(qnet)
score = test(qnet, testloader, cuda=False)
print('Accuracy of the fused and quantized network on the test images: {}% - INT8'.format(score))
```

```
QConfig(activation=functools.partial(<class 'torch.ao.quantization.
observer.MovingAverageMinMaxObserver'>, reduce_range=True){},
weight=functools.partial(<class 'torch.ao.quantization.observer.
MovingAverageMinMaxObserver'>, dtype=torch.qint8, qscheme=torch.per_tensor_
symmetric){})
Post Training Quantization Prepare: Inserting Observers

 Conv1: After observer insertion

 ConvReLU2d(
  (0): Conv2d(1, 6, kernel_size=(5, 5), stride=(1, 1), bias=False)
  (1): ReLU()
  (activation_post_process): MovingAverageMinMaxObserver(min_val=inf,
  max_val=-inf)
)
/usr/local/lib/python3.7/dist-packages/torch/ao/quantization/observer.
py:178: UserWarning: Please use quant_min and quant_max to specify the
```

range for observers.   reduce_range will be deprecated in
                                             a future release of PyTorch.
  reduce_range will be deprecated in a future release of PyTorch."
Post Training Quantization: Calibration done
Post Training Quantization: Convert done

 Conv1: After fusion and quantization

 QuantizedConvReLU2d(1, 6, kernel_size=(5, 5), stride=(1, 1),
scale=0.06884118169546127, zero_point=0, bias=False)
Size of model after quantization
Size (MB): 0.049714
Accuracy of the fused and quantized network on the test images:
98.6% - INT8

接下来，将观察器更改为 HistogramObserver。

In [40]:

qnet = Net(q=True)
load_model(qnet, net)
fuse_modules(qnet)

In [41]:

qnet.qconfig = torch.quantization.get_default_qconfig('fbgemm')
print(qnet.qconfig)

torch.quantization.prepare(qnet, inplace=True)
test(qnet, trainloader, cuda=False)
torch.quantization.convert(qnet, inplace=True)
print("Size of model after quantization")
print_size_of_model(qnet)
QConfig(activation=functools.partial(<class 'torch.ao.quantization.
observer.HistogramObserver'>, reduce_range=True){}, weight=functools.
partial(<class 'torch.ao.quantization.observer.PerChannelMinMaxObserver'>,
dtype=torch.qint8, qscheme=torch.per_channel_symmetric){})
Size of model after quantization
Size (MB): 0.055572

经过量化处理后，模型的大小得以减小，同时模型的准确性并未受到很大影响。

In [42]:

score = test(qnet, testloader, cuda=False)

```
print('Accuracy of the fused and quantized network on the test images:
{}% - INT8'.format(score))
Accuracy of the fused and quantized network on the test images:
98.58% - INT8
```

接下来,我们将运用量化感知训练(Quantization Aware Training,QAT)技术。该技术通过将数字四舍五入,对所有权重和激活值进行"错误"的量化,使其转变为浮点数。为确保模型精度,需要在量化配置上进行一些额外的调整。在重新训练之后,模型的准确率也得到了提升,达到 98.69%。

```
In [43]:
qnet = Net(q=True)
fuse_modules(qnet)
qnet.qconfig = torch.quantization.get_default_qat_qconfig('fbgemm')
torch.quantization.prepare_qat(qnet, inplace=True)
print('\n Conv1: After fusion and quantization \n\n', qnet.conv1)
qnet=qnet
Conv1: After fusion and quantization

ConvReLU2d(
  1, 6, kernel_size=(5, 5), stride=(1, 1), bias=False
  (weight_fake_quant): FusedMovingAvgObsFakeQuantize(
    fake_quant_enabled=tensor([1]), observer_enabled=tensor([1]),
    scale=tensor([1.]), zero_point=tensor([0], dtype=torch.int32),
    dtype=torch.qint8, quant_min=-128, quant_max=127, qscheme=torch.
    per_channel_symmetric, reduce_range=False
    (activation_post_process): MovingAveragePerChannelMinMaxObserver(min_
    val=tensor([]), max_val=tensor([]))
  )
  (activation_post_process): FusedMovingAvgObsFakeQuantize(
    fake_quant_enabled=tensor([1]), observer_enabled=tensor([1]),
    scale=tensor([1.]), zero_point=tensor([0], dtype=torch.int32),
    dtype=torch.quint8, quant_min=0, quant_max=127, qscheme=torch.
    per_tensor_affine, reduce_range=True
    (activation_post_process): MovingAverageMinMaxObserver(min_val=inf,
max_val=-inf)
  )
)
```

In[44]:
```
train(qnet, trainloader, cuda=False)
```

```
[15,   301]  loss 0.024571 (0.050655) train_acc 100.000000 (98.421927)
[15,   401]  loss 0.083018 (0.052274) train_acc 96.875000 (98.394638)
[15,   501]  loss 0.137679 (0.053624) train_acc 96.875000 (98.334581)
[15,   601]  loss 0.067262 (0.053238) train_acc 98.437500 (98.328307)
[15,   701]  loss 0.025749 (0.053818) train_acc 100.000000 (98.334968)
[15,   801]  loss 0.100811 (0.054027) train_acc 96.875000 (98.341916)
[15,   901]  loss 0.040471 (0.053821) train_acc 98.437500 (98.336917)
[16,     1]  loss 0.024462 (0.024462) train_acc 100.000000 (100.000000)
[16,   101]  loss 0.047599 (0.050869) train_acc 96.875000 (98.452970)
```

In[45]:
```
qnet = qnet.cpu()
torch.quantization.convert(qnet, inplace=True)
print("Size of model after quantization")
print_size_of_model(qnet)

score = test(qnet, testloader, cuda=False)
print('Accuracy of the fused and quantized network (trained quantized) on the test images: {}% - INT8'.format(score))
```

```
Size of model after quantization
Size (MB): 0.055572
Accuracy of the fused and quantized network (trained quantized) on the test images: 98.69% - INT8
```

经过应用融合和量化处理后，模型尺寸成功缩减至0.055MB，同时精度略有降低。

In[46]:
```
qnet = Net(q=True)
fuse_modules(qnet)
qnet.qconfig = torch.quantization.get_default_qat_qconfig('fbgemm')
torch.quantization.prepare_qat(qnet, inplace=True)
qnet = qnet
train(qnet, trainloader, cuda=False, q=True)
qnet = qnet.cpu()
torch.quantization.convert(qnet, inplace=True)
```

```
print("Size of model after quantization")
print_size_of_model(qnet)

score = test(qnet, testloader, cuda=False)
print('Accuracy of the fused and quantized network (trained quantized) on
the test images: {}% - INT8'.format(score))
[15,   601] loss 0.070189 (0.060619) train_acc 98.437500 (98.128120)
[15,   701] loss 0.275282 (0.060933) train_acc 93.750000 (98.123217)
[15,   801] loss 0.029783 (0.060075) train_acc 100.000000 (98.172207)
[15,   901] loss 0.028618 (0.059222) train_acc 98.437500 (98.186043)
[16,     1] loss 0.046160 (0.046160) train_acc 98.437500 (98.437500)
[16,   101] loss 0.110157 (0.069398) train_acc 96.875000 (98.004332)
[16,   201] loss 0.157249 (0.063327) train_acc 95.312500 (98.173197)
[16,   301] loss 0.013502 (0.059892) train_acc 100.000000 (98.183140)

Finished Training
Size of model after quantization
Size (MB): 0.055572
Accuracy of the fused and quantized network (trained quantized) on the test
images: 98.53% - INT8
```

我们可以得出以下结论：在此特定示例中，经过量化处理后，模型的大小显著减小，同时模型的准确性并未明显降低。然而，这一情况并非普遍适用。在应用量化过程中，需逐步进行并密切关注准确性变化。量化应用的主要目标是提高推理生成速度，若准确性受到损害，则量化应用的价值将大打折扣。因此，量化感知训练通常能够达到接近浮点精度的准确性。

小结

在本章中，我们探讨了在 GPU 环境下运用分布式 PyTorch 的应用方法，并研究了以并行方式处理模型训练的技巧。此外，还了解到了将大型深度学习模型对象压缩为较小尺寸的策略，同时保持模型准确率。这种压缩量化对于提高深度学习模型的推理生成速度具有关键作用。本章详细讨论了各种压缩量化方法。在下一章中，将阐述应用于图像和音频数据的数据增强技术，以及利用 PyTorch 进行特征工程和提取相关特征的方法。

第 9 章

图像和音频的数据增强、特征工程和提取

在音频分类任务中，我们希望深度学习算法能够对声音进行学习，并准确预测其所属类别。同样地，在图像分类任务中，我们期望深度学习模型能够理解图像，从中提取特定模式，并将新图像自动归类至训练算法已学习的相应类别中。在处理声音分类时，我们通常将音频文件作为输入，并将其转换为一种特殊形式，即频谱图。频谱图构成了一个高维数据空间，可以通过应用卷积神经网络模型对其进行降维处理。众所周知，卷积神经网络模型的最后一层是一个神经网络，也被称为全连接层，该层通常用作分类器。

在本章中，将探讨如何运用 PyTorch 库实现音频处理与图像处理任务中最常用的步骤。

秘籍 9-1　音频处理中的频谱图

问题

在使用频谱图进行训练时，如何增强音频数据？

解决方案

原始音频文件可采用 mp4、mp3 或 wav 格式，然而此类格式无法直接应用于模型训练。这是因为模型训练过程需要数据以结构化表格形式出现，因此核心问题在于如何将音频文件转换为表格格式。

编程实战

在将深度学习模型应用于音频数据转换的过程中，首先，根据运行环境是 CPU 还是 GPU，将音频文件从 wav 格式加载至内存。接下来，将 wav 文件转换为立体声格式。然后进行时移音频增强，并把音频文件转换为频谱图。通常情况下，由于立体声的特性，音频文件含有两个通道，然而在部分场合，音频片段可能仅来自一个通道。因此，在开始训练模型之前，需进行标准化处理，以确保输入数据的标准化。

在 PyTorch 框架中，Torchaudio 模块包含两个子组件，用于处理音频数据，它们分别是：

- Torchaudio.functional：该组件提供了一系列函数，用于对音频数据进行操作和处理。这些函数可以执行各种任务，例如音频特征提取、音频生成以及音频转换等。
- Torchaudio.transforms：该组件包含了一些用于音频数据预处理和后处理的变换方法。这些变换方法可以用于音频数据的归一化、剪切、填充以及重新采样等操作，以确保音频数据在不同的处理流程中能够保持一致性。

这两个组件都为开发者提供了灵活且高效的方式来处理音频数据，从而使得音频数据处理任务变得更加简单和高效。

In[1]:
```
import torch
import torchaudio
import torchaudio.functional as F
import torchaudio.transforms as T

print(torch.__version__)
```

```python
print(torchaudio.__version__)
from IPython.display import Audio
```

In[2]:
```python
import librosa
import matplotlib.pyplot as plt
from torchaudio.utils import download_asset

torch.random.manual_seed(0)
SAMPLE_SPEECH = download_asset("YOURSAMPLE_AUDIO_DATA.wav")
```

以下是将音频文件转换为波形图显示的函数示例。

In[3]:
```python
def plot_waveform(waveform, sr, title="Waveform"):
    waveform = waveform.numpy()

    num_channels, num_frames = waveform.shape
    time_axis = torch.arange(0, num_frames) / sr

    figure, axes = plt.subplots(num_channels, 1)
    axes.plot(time_axis, waveform[0], linewidth=1)
    axes.grid(True)
    figure.suptitle(title)
    plt.show(block=False)
```

以下是将输入数据转换为频谱图的函数示例。

In[4]:
```python
def plot_spectrogram(specgram, title=None, ylabel="freq_bin"):
    fig, axs = plt.subplots(1, 1)
    axs.set_title(title or "Spectrogram (db)")
    axs.set_ylabel(ylabel)
    axs.set_xlabel("frame")
    im = axs.imshow(librosa.power_to_db(specgram), origin="lower",
    aspect="auto")
    fig.colorbar(im, ax=axs)
    plt.show(block=False)
```

在 Torchaudio 中的 functional 模块被称为"无状态",这是因为该模块可以作为独立的函数实现,无需依赖任何外部状态。这种设计使得 functional 模块具备更高的灵活性和可重用性,能在各种音频处理任务中反复应用,同时避免了状态保存与重置的问题。

秘籍 9-2　安装 Torchaudio

问题

如何使用 PyTorch 安装 Torchaudio？

解决方案

在安装 Torchaudio 模块时，需注意其特定版本的需求。由于默认版本存在问题，因此在使用 PIP 进行安装时，务必指定相应版本。完成安装后，需重新启动 Google Colab 运行环境或 Python Jupyter 环境。

编程实战

在成功安装并导入 Torchaudio 模块后，PyTorch 中与音频处理相关的功能将能够正常运行。

```
In[5]:
pip install torchaudio==0.4.0
Looking in indexes: https://pypi.org/simple, https://us-python.pkg.dev/
colab-wheels/public/simple/
Collecting torchaudio==0.4.0
  Downloading torchaudio-0.4.0-cp37-cp37m-manylinux1_x86_64.whl (3.1 MB)
     |████████████████████████████████| 3.1 MB 8.6 MB/s
Collecting torch==1.4.0
  Downloading torch-1.4.0-cp37-cp37m-manylinux1_x86_64.whl (753.4 MB)
     |████████████████████████████████| 753.4 MB 6.4 kB/s
Installing collected packages: torch, torchaudio
```

```
  Attempting uninstall: torch
    Found existing installation: torch 1.7.0+cpu
    Uninstalling torch-1.7.0+cpu:
      Successfully uninstalled torch-1.7.0+cpu
  Attempting uninstall: torchaudio
    Found existing installation: torchaudio 0.7.0
    Uninstalling torchaudio-0.7.0:
      Successfully uninstalled torchaudio-0.7.0
ERROR: pip's dependency resolver does not currently take into account
all the packages that are installed. This behaviour is the source of the
following dependency conflicts.
torchvision 0.8.1+cpu requires torch==1.7.0, but you have torch 1.4.0 which
is incompatible.
torchtext 0.13.1 requires torch==1.12.1, but you have torch 1.4.0 which is
incompatible.
fastai 2.7.9 requires torch<1.14,>=1.7, but you have torch 1.4.0 which is
incompatible.
fastai 2.7.9 requires torchvision>=0.8.2, but you have torchvision
0.8.1+cpu which is incompatible.
Successfully installed torch-1.4.0 torchaudio-0.4.0
```

**WARNING: The following packages were previously imported in this runtime:
  [torch]**

You must restart the runtime in order to use newly installed versions.

Please note the last line as restart is mandatory. If that does not happen, you may get an error.

In [6]:
```
import torchaudio
import torchaudio.functional as F
import torchaudio.transforms as T
```

In [7]:
```
print(torch.__version__)
print(torchaudio.__version__)
```

秘籍 9-3　将音频文件加载到 PyTorch 中

问题

如何使用 PyTorch 将数据加载到 Torchaudio 中？

解决方案

Torchaudio 模块内置了用于训练深度学习模型的数据集，然而，在实际应用中，加载其他类型文件作为原始数据并实施相应转换的过程，才是真正的难点部分。

编程实战

以下代码脚本解释了如何从内置库和本地目录加载数据。

```
In[8]:
yesno_data = torchaudio.datasets.YESNO('.', download=True)
data_loader = torch.utils.data.DataLoader(yesno_data,
                                          batch_size=1,
                                          shuffle=True,
                                          num_workers=2)
```

要以外部 URL 方式读取音频数据，请按照以下步骤进行。样例数据的波形图如图 9-1 所示。

```
In[9]:
audio_url = "https://pytorch.org/tutorials/_static/img/steam-train-whistle-daniel_simon-converted-from-mp3.wav"
request_url = requests.get(audio_url)
In[10]:
with open('steam-train-whistle-daniel_simon-converted-from-mp3.wav', 'wb') as file:
```

```
        file.write(request_url.content)
In[11]:
audio_file = "steam-train-whistle-daniel_simon-converted-from-mp3.wav"
data_waveform, rate_of_sample = torchaudio.load(audio_file)
In[12]:
print("This is the shape of the waveform: {}".format(data_waveform.size()))
print("This is the output for Sample rate of the waveform: {}".format(rate_of_sample))
This is the shape of the waveform: torch.Size([2, 276858])
This is the output for Sample rate of the waveform: 44100
In[13]:
plt.figure()
plt.plot(data_waveform.t().numpy())
```

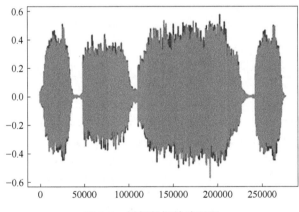

图 9-1　样例数据的波形图

秘籍 9-4　安装用于音频的 Librosa 库

问题

如何使用 PyTorch 安装 Librosa 以进行声音数据转换？

解决方案

从音频与声音文件中提取特征并实施转换,需依赖 Librosa 这一 Python 工具包所提供的函数。Librosa 是一款功能丰富的库,用于音乐和声音分析。

编程实战

首先,创建一个新的 Jupyter notebook 或 Colab notebook。

In[14]:
```
!pip install librosa
Looking in indexes: https://pypi.org/simple, https://us-python.pkg.dev/colab-wheels/public/simple/
Requirement already satisfied: librosa in /usr/local/lib/python3.7/dist-packages (0.8.1)
Requirement already satisfied: decorator>=3.0.0 in /usr/local/lib/python3.7/dist-packages (from librosa) (4.4.2)
Requirement already satisfied: scipy>=1.0.0 in /usr/local/lib/python3.7/dist-packages (from librosa) (1.7.3)
Requirement already satisfied: numba>=0.43.0 in /usr/local/lib/python3.7/dist-packages (from librosa) (0.56.0)
Requirement already satisfied: packaging>=20.0 in /usr/local/lib/python3.7/dist-packages (from librosa) (21.3)
Requirement already satisfied: audioread>=2.0.0 in /usr/local/lib/python3.7/dist-packages (from librosa) (3.0.0)
Requirement already satisfied: resampy>=0.2.2 in /usr/local/lib/python3.7/dist-packages (from librosa) (0.4.0)
Requirement already satisfied: soundfile>=0.10.2 in /usr/local/lib/python3.7/dist-packages (from librosa) (0.10.3.post1)
Re
```

In[15]:
```
from IPython.display import Audio
import librosa
import matplotlib.pyplot as plt
from torchaudio.utils import download_asset
```

```
torch.random.manual_seed(0)
SAMPLE_SPEECH = download_asset("/content/waves_yesno/0_0_0_0_1_1_1_1.wav")
```

加载声音文件并实现可视化，可以参考以下代码示例。原始波形图如图 9-2 所示。

In [16]:

```
SPEECH_WAVEFORM, SAMPLE_RATE = torchaudio.load(SAMPLE_SPEECH)

plot_waveform(SPEECH_WAVEFORM, SAMPLE_RATE, title="Original waveform")
Audio(SPEECH_WAVEFORM.numpy(), rate=SAMPLE_RATE)
```

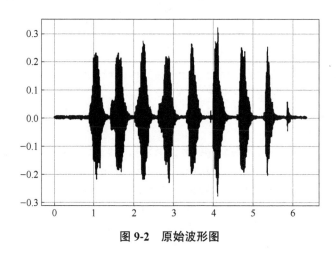

图 9-2 原始波形图

秘籍 9-5 频谱图变换

问题

如何从声音文件创建频谱图？

解决方案

频谱图也称为声谱图，可理解为对频率谱在时间维度上的变化进行可视化呈现的一种图形化表示方式。

编程实战

在音频数据应用于深度学习模型过程中，涉及多个环节，其中一项便是频谱图处理。如图 9-3 所示。

In[17]:
```
import torchaudio.transforms as T
n_fft = 1024
win_length = None
hop_length = 512
# 定义频谱转换
spectrogram = T.Spectrogram(
    n_fft=n_fft,
    win_length=win_length,
    hop_length=hop_length,
    center=True,
    pad_mode="reflect",
    power=2.0,
)
```

In[18]:
```
# 执行频谱转换
spec = spectrogram(SPEECH_WAVEFORM)
```

In[19]:
```
plot_spectrogram(spec[0], title="torchaudio")
```

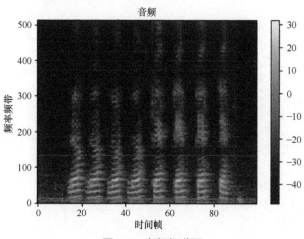

图 9-3　音频频谱图

秘籍 9-6　Griffin-Lim 变换

问题

什么是 Griffin-Lim 变换，应当如何进行变换？

解决方案

Griffin-Lim 算法（GLA）是一种常用的声谱图到波形的逆转换算法。该算法通过迭代投影的方式，尝试将初始的随机相位谱图与幅度谱图相结合，从而逐渐恢复出原始的波形。对于信号来说，保持幅度一致的连贯声谱图是必需的，因此在从声谱图恢复波形时，GLA 转换是数据增强的一种必要手段。

编程实战

应用 GLA 的过程可通过以下代码实现。使用 GLA 重建的波形图如图 9-4 所示。

In [20]:
```
# 要从频谱图中恢复波形，可以使用 GLA
```
In [21]:
```
import torchaudio.transforms as T

torch.random.manual_seed(0)

n_fft = 1024
win_length = None
hop_length = 512

spec = T.Spectrogram(
    n_fft=n_fft,
    win_length=win_length,
    hop_length=hop_length,
)(SPEECH_WAVEFORM)
```

In [22]:
```
griffin_lim = T.GriffinLim(
    n_fft=n_fft,
    win_length=win_length,
    hop_length=hop_length,
)
```
In [23]:
```
reconstructed_waveform = griffin_lim(spec)
```
In [24]:
```
plot_waveform(reconstructed_waveform, SAMPLE_RATE, title="Reconstructed")
Audio(reconstructed_waveform, rate=SAMPLE_RATE)
```

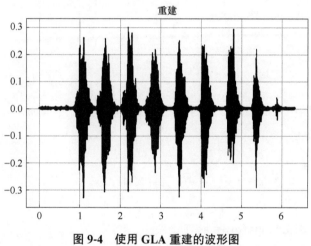

图 9-4 使用 GLA 重建的波形图

秘籍 9-7 使用滤波器组进行梅尔尺度转换

问题

滤波器组在将频率频带转换为梅尔尺度频带方面的应用是怎样的？

解决方案

为了将频率频带转换为梅尔尺度频带，Torchaudio 功能模块提供了一个滤波器组。此过程无需输入音频特征。滤波器组可以被定义为将连续的频率响应离散化为不同的频带的方法。滤波器组的类型取决于实际应用场景。梅尔滤波器组是众多滤波器类型中的一种。

频率频带指的是将频率范围划分为等宽的离散区间,例如从 0Hz 到采样率的一半,通过将该范围划分为一定数量的区间(频带),每个区间代表一定的频率范围。频率频带通常与傅里叶变换相关,用于表示音频信号在频域上的能量分布。而梅尔尺度频带是一种非线性的频率刻度,更符合人类听觉系统的感知特性。梅尔尺度是根据人类听觉感知的频率间隔变化而设计的。梅尔尺度频带通过将频率范围转换为梅尔尺度上的离散区间,以更好地捕捉音频信号中不同频率之间的感知差异。在梅尔尺度上,较低频率的区间较宽,较高频率的区间较窄,以反映人耳对不同频率的感知敏感度。

编程实战

我们可以调用 Torchaudio 中的梅尔尺度转换函数。Torchaudio 中的梅尔滤波器组如图 9-5 所示。

In [25]:
```
# 用于将频率频带转换为梅尔尺度频带的滤波器组
```
In [26]:
```
import torchaudio.transforms as T
n_fft = 255
n_mels = 61
sample_rate = 5000

mel_filters = T.melscale_fbanks(
    int(n_fft // 2 + 1),
    n_mels=n_mels,
    f_min=0.0,
    f_max=sample_rate / 2.0,
    sample_rate=sample_rate,
    norm="slaney",
)
```
In [27]:
```
plot_fbank(mel_filters, "Mel Filter Bank - torchaudio")
```

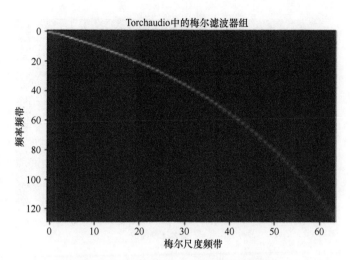

图 9-5　Torchaudio 中的梅尔滤波器组

图 9-6 展示了 Librosa 库中的梅尔滤波器组,其应用效果与 Torchaudio 中的滤波器相似。通过使用 Librosa 库,可以实现相同的滤波功能。

In[28]:

Librosa 库中的梅尔滤波器组

In[29]:

```
mel_filters_librosa = librosa.filters.mel(
    sr=sample_rate,
    n_fft=n_fft,
    n_mels=n_mels,
    fmin=0.0,
    fmax=sample_rate / 2.0,
    norm="slaney",
    htk=True,
).T
```

In[30]:

```
plot_fbank(mel_filters_librosa, "Mel Filter Bank - librosa")

mse = torch.square(mel_filters - mel_filters_librosa).mean().item()
print("Mean Square Difference: ", mse)
```

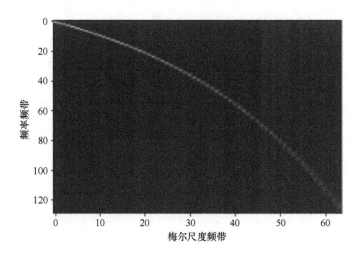

图 9-6　Librosa 库中的梅尔滤波器组

秘籍 9-8　Librosa 的梅尔尺度转换与 Torchaudio 版本对比

问题

如何比较 Librosa 库中的梅尔尺度转换和 PyTorch 中的 Torchaudio 库？

解决方案

Torchaudio 模块为开发者提供了便捷的音频处理工具，其中包括生成梅尔频谱图的功能。在构建梅尔频谱图时，需设定采样率、窗口长度、跳跃长度、填充以及功率等参数。通过传入一个音频样本的波形数据，即可生成相应的频谱图。图 9-7 所示为梅尔频谱图。

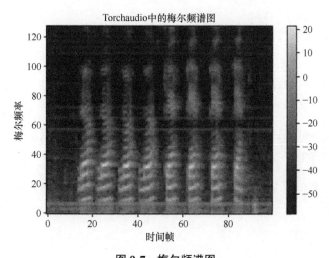

图 9-7 梅尔频谱图

编程实战

可以使用 Librosa 和 Torchaudio 库来实现音频数据转换。

In[31]:
生成频谱图并执行梅尔尺度转换

In[32]:
```
import torchaudio.transforms as T
n_fft = 1024
win_length = None
hop_length = 512
n_mels = 128

mel_spectrogram = T.MelSpectrogram(
    sample_rate=sample_rate,
    n_fft=n_fft,
    win_length=win_length,
hop_length=hop_length,
center=True,
pad_mode="reflect",
power=2.0,
norm="slaney",
onesided=True,
```

```python
    n_mels=n_mels,
    mel_scale="htk",
)
melspec = mel_spectrogram(SPEECH_WAVEFORM)
```

In [33]:
```python
plot_spectrogram(melspec[0], title="MelSpectrogram - torchaudio",
ylabel="mel freq")
```

接下来，我们可以使用Librosa生成梅尔尺度频谱图。图9-8所示为转换后的梅尔频谱图。

In [34]:
```python
# 使用Librosa生成梅尔尺度频谱图的等效方法
```

In [35]:
```python
melspec_librosa = librosa.feature.melspectrogram(
    y=SPEECH_WAVEFORM.numpy()[0],
    sr=sample_rate,
    n_fft=n_fft,
    hop_length=hop_length,
    win_length=win_length,
    center=True,
    pad_mode="reflect",
    power=2.0,
    n_mels=n_mels,
    norm="slaney",
    htk=True,
)
```

In [36]:
```python
plot_spectrogram(melspec_librosa, title="MelSpectrogram - librosa",
ylabel="mel freq")

mse = torch.square(melspec - melspec_librosa).mean().item()
print("Mean Square Difference: ", mse)
```

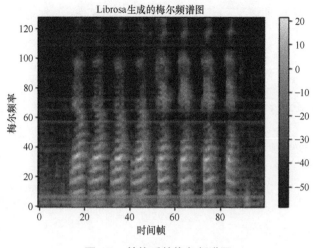

图 9-8 转换后的梅尔频谱图

秘籍 9-9　使用 Librosa 和 Torchaudio 进行 MFCC 和 LFCC

问题

如何将梅尔频率倒谱系数（MFCC）和线性频率倒谱系数（LFCC）应用于增强语音数据？

解决方案

Torchaudio 模块提供了一系列数据增强策略，根据所选算法及应用于语音数据的方式，这些策略的具体实现可能有所不同。例如，若要应用线性模型，如高斯混合模型（GMM），则需首先通过离散余弦变换（DCT）计算 MFCC。

编程实战

MFCC 作为一种线性模型压缩表示方法，适用于有限数据集的场景。然而，在处理大规模数据集且涉及分类任务的情景中，卷积神经网络具有显著优势。相较之下，MFCC 在这种情况下表现更为出色。MFCC 频谱图如图 9-9 所示。

In[37]:
MFCC

In[38]:
```
import torchaudio.transforms as T
n_fft = 2048
win_length = None
hop_length = 512
n_mels = 256
n_mfcc = 256

mfcc_transform = T.MFCC(
    sample_rate=sample_rate,
    n_mfcc=n_mfcc,
    melkwargs={
        "n_fft": n_fft,
        "n_mels": n_mels,
        "hop_length": hop_length,
        "mel_scale": "htk",
    },
)

mfcc = mfcc_transform(SPEECH_WAVEFORM)
```

In[39]:
```
plot_spectrogram(mfcc[0])
```

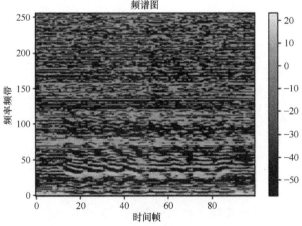

图 9-9　MFCC 频谱图

使用 Librosa 可以实现类似的频谱图表示（见图 9-10）。

In[40]:

```
melspec = librosa.feature.melspectrogram(
    y=SPEECH_WAVEFORM.numpy()[0],
    sr=sample_rate,
    n_fft=n_fft,
    win_length=win_length,
    hop_length=hop_length,
    n_mels=n_mels,
    htk=True,
    norm=None,
)

mfcc_librosa = librosa.feature.mfcc(
    S=librosa.core.spectrum.power_to_db(melspec),
    n_mfcc=n_mfcc,
    dct_type=2,
    norm="ortho",
)
```

In[41]:

```
plot_spectrogram(mfcc_librosa)

mse = torch.square(mfcc - mfcc_librosa).mean().item()
print("Mean Square Difference: ", mse)
```

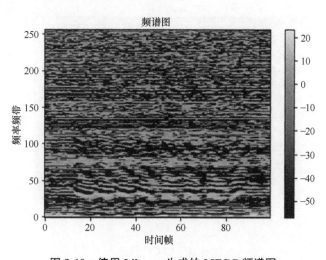

图 9-10　使用 Librosa 生成的 MFCC 频谱图

以下代码示例展示了如何实现 LFCC 转换技术。LFCC 频谱图如图 9-11 所示。

In [42]:

#LFCC

In [43]:
```
import torchaudio.transforms as T
n_fft = 2048
win_length = None
hop_length = 512
n_lfcc = 256

lfcc_transform = T.LFCC(
    sample_rate=sample_rate,
    n_lfcc=n_lfcc,
    speckwargs={
        "n_fft": n_fft,
        "win_length": win_length,
        "hop_length": hop_length,
    },
)

lfcc = lfcc_transform(SPEECH_WAVEFORM)
plot_spectrogram(lfcc[0])
```

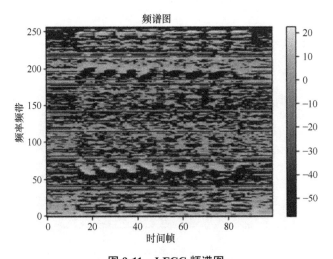

图 9-11　LFCC 频谱图

使用 Librosa 库，我们同样可以实现相同的 LFCC 频谱图。此处不再演示。

秘籍 9-10 图像数据增强

问题

如何使用 PyTorch 通过应用图像变换来扩充图像数据？

解决方案

接下来，我们将运用 CIFAR10 数据集，探讨如何利用 Transforms 和 Compose 函数对图像数据进行扩充。图 9-12~图 9-14 展示了 CIFAR10 数据集的部分样例图像。

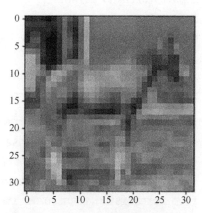

图 9-12 CIFAR10 数据集样例图像一

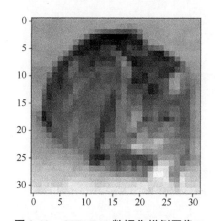

图 9-13 CIFAR10 数据集样例图像二

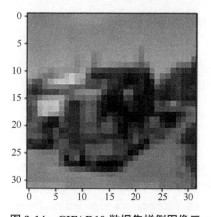

图 9-14 CIFAR10 数据集样例图像三

编程实战

以下代码片段展示了数据扩充滤波器的实现及其应用。

In[44]:
```python
import os
import sys
import random
import tempfile
import torch
import torch.distributed as dist
import torch.nn as nn
import torch.optim as optim
import torch.multiprocessing as mp
import torchvision
import torchvision.datasets as datasets
import torchvision.transforms as transforms
# 导入与数据集和数据加载器相关的包
from torchvision import datasets
from torchvision.transforms import ToTensor
from torch.utils.data import DataLoader
from torchvision.transforms import Compose, Grayscale
```

In[45]:
```python
# 从CIFAR10数据集中下载并加载图像
cifar10_data = datasets.CIFAR10(
    root="data",  # 图像将存储的路径
    download=True,  # 所有图片都将被下载下来
    transform=ToTensor()  # 将图像转换为张量
)
# 打印已加载数据集中的样本数量
print(f"Number of samples: {len(cifar10_data)}")
print(f"Class names: {cifar10_data.classes}")
```
Files already downloaded and verified
Number of samples: 50000
Class names: ['airplane', 'automobile', 'bird', 'cat', 'deer', 'dog', 'frog', 'horse', 'ship', 'truck']

In[46]:
```python
# 选择随机样本
random.seed(2021)
```

```python
image, label = cifar10_data[random.randint(0, len(cifar10_data))]
print(f"Label: {cifar10_data.classes[label]}")
print(f"Image size: {image.shape}")
Label: horse
Image size: torch.Size([3, 32, 32])
```

In[47]:
```python
import matplotlib.pyplot as plt
plt.imshow(image.permute(1, 2, 0))
plt.show()
```

In[48]:
```python
data = datasets.CIFAR10(root="data", download=True,
                        transform=Compose([ToTensor(), Grayscale()]))
# 显示一张随机的灰度图像
image, label = data[random.randint(0, len(data))]
plt.imshow(image.squeeze(), cmap="gray")
plt.show()
```

In[49]:
```python
# 加载训练样本
training_data = datasets.CIFAR10(
    root="data",
    train=True,
    download=True,
    transform=ToTensor()
)
# 加载测试样本
test_data = datasets.CIFAR10(
    root="data",
    train=False,
    download=True,
    transform=ToTensor()
)
```

In[50]:
```python
# 创建数据加载器
train_dataloader = DataLoader(training_data, batch_size=64, shuffle=True)
```

```
test_dataloader = DataLoader(test_data, batch_size=64, shuffle=True)
```
In[51]:
加载下一批次数据
```
batch_images, batch_labels = next(iter(train_dataloader))
print('Batch size:', batch_images.shape)
```
显示批次中的第一张图片
```
plt.imshow(batch_images[0].permute(1, 2, 0))
plt.show()
```
In[52]:
在数据集加载过程中，应用数据转换是一项至关重要的功能，这包括但不限于颜色空间的转换、数据归一化、图像裁剪以及旋转等操作。torchvision.transforms 包提供了众多预定义的转换函数，以满足用户的不同需求。此外，用户还可以利用 Compose 转换功能，将多个转换操作组合在一起，实现更为复杂和灵活的数据预处理流程

In[53]:
```
transform = transforms.Compose(
    [transforms.ToTensor(),
     transforms.Normalize((0.5,), (0.5,))])
```

在上述代码中，我们运用了归一化作为数据增强手段，同样，也可以在原始图像上实施随机裁剪和随机水平翻转等操作。关于这些额外的数据增强脚本，可以参考 PyTorch 官方文档页面。

小结

在本章中，我们探讨了如何运用数据增强技术以提高音频和图像模型的性能，涵盖了音频的波形转换、图像的滤波和增强等方法。接下来的一章将介绍 Skorch 库和 Captum 工具。使用 Skorch 库可以轻松地应用 Scikit-learn 函数和 API，如管道、网格搜索和交叉验证等。Captum 工具则可以在深度学习模型上实现模型可解释性。

第 10 章

PyTorch 模型可解释性和 Skorch

模型的可解释性是一个广泛关注的问题，其关乎模型的可信度，乃至人工智能的普及程度。当使用者能够理解和掌握模型的决策及预测过程，他们将更易于建立对模型的信任并加以应用。为此，本章将阐述一个名为 Captum 的新框架，该框架提供了一系列算法，有助于我们解释神经网络模型的预测结果、模型参数以及各层网络的作用。通过运用 Captum，可以更深入地理解模型是如何做出决策的，进而提升模型的透明度和可信度。此外，本章还将介绍另一个名为 Skorch 的框架，该框架是一个与 PyTorch 深度学习框架兼容的 Sklearn 库。Skorch 为用户带来了类似 Sklearn 的便捷体验，使用户能够轻松地训练神经网络模型、执行网格搜索并寻找到最佳的超参数配置。

Captum 库内置了多种解释性方法，为基于 PyTorch 框架的深度学习模型提供了强大的解释工具。在神经网络解释过程中，运用这些方法有助于更清晰地揭示特征重要性、主导层识别以及主导神经元识别等关键信息。Captum 库提供三种归因算法，使得获取这些信息更为便捷且精确。

- 主要归因：帮助解释特征重要性。
- 层归因：帮助识别给定层中每个神经元对模型输出的贡献。
- 神经元归因：帮助识别每个输入特征对神经元激活的影响。

在本章中，我们将使用 Skorch 和 Captum 来实现模型可解释性，同时学习与 Sklearn 兼容性中最常用的步骤。

秘籍 10-1　安装 Captum 库

问题

如何安装 Captum 库？

解决方案

有两种安装 Captum 库的方式可以选择：使用 conda 或者 pip 命令。

编程实战

以下命令行脚本可用于安装 Captum 库。

In[1]:
```
conda install captum -c pytorch
```
或
```
pip install captum
```
Out[1]:
```
Looking in indexes: https://pypi.org/simple, https://us-python.pkg.dev/
colab-wheels/public/simple/
Collecting captum
  Downloading captum-0.5.0-py3-none-any.whl (1.4 MB)
     |████████████████████████████████| 1.4 MB 29.2 MB/s
Requirement already satisfied: numpy in /usr/local/lib/python3.7/dist-
packages (from captum) (1.21.6)
Requirement already satisfied: matplotlib in /usr/local/lib/python3.7/dist-
packages (from captum) (3.2.2)
Requirement already satisfied: torch>=1.6 in /usr/local/lib/python3.7/dist-
packages (from captum) (1.12.1+cu113)
Requirement already satisfied: typing-extensions in /usr/local/lib/
python3.7/dist-packages (from torch>=1.6->captum) (4.1.1)
```

```
Requirement already satisfied: python-dateutil>=2.1 in /usr/local/lib/
python3.7/dist-packages (from matplotlib->captum) (2.8.2)
Requirement already satisfied: pyparsing!=2.0.4,!=2.1.2,!=2.1.6,>=2.0.1 in
/usr/local/lib/python3.7/dist-packages (from matplotlib->captum) (3.0.9)
Requirement already satisfied: cycler>=0.10 in /usr/local/lib/python3.7/
dist-packages (from matplotlib->captum) (0.11.0)
Requirement already satisfied: kiwisolver>=1.0.1 in /usr/local/lib/
python3.7/dist-packages (from matplotlib->captum) (1.4.4)
Requirement already satisfied: six>=1.5 in /usr/local/lib/python3.7/dist-
packages (from python-dateutil>=2.1->matplotlib->captum) (1.15.0)
Installing collected packages: captum
Successfully installed captum-0.5.0
```

使用 Anaconda 进行安装是一种既安全又便捷的方式。在安装过程中，需关注 Python 的版本依赖和 PyTorch 的版本依赖。为确保安装顺利进行，建议使用不低于 3.6 版本的 Python 以及不低于 1.2 版本的 PyTorch。

秘籍 10-2　主要归因：深度学习模型的特征重要性

问题

如何使用 Captum 来实现主要归因？

解决方案

主要归因层提供了集成梯度（Integrated Gradients，IG）、梯度夏普利加法解释（Shapley Additive Explanations，SHAP）以及显著性等方法，以更好地解释模型。在此，我们将使用常见的 titanic.csv 数据集为例，运用 PyTorch 构建一个分类模型，并运用 IG 等多种归因分析方法，对模型进行主要归因分析。该数据集包含了关于灾难中幸存者和未幸存者的特征和表示，将以输出列"survived"为基础，设计一个分类模型。

编程实战

以下代码展示了如何实现归因层，其中 IG 代表对输入路径的梯度积分。

In[2]:
```
# 初始导入的包
import numpy as np

import torch

from captum.attr import IntegratedGradients
from captum.attr import LayerConductance
from captum.attr import NeuronConductance

import matplotlib
import matplotlib.pyplot as plt
%matplotlib inline

from scipy import stats
import pandas as pd
```

In[3]:
```
dataset_path = "https://raw.githubusercontent.com/pradmishra1/PublicDatasets/main/titanic.csv"
```

In[4]:
```
titanic_data = pd.read_csv(dataset_path)
```

In[5]:
```
del titanic_data['Unnamed: 0']
```

In[6]:
```
del titanic_data['PassengerId']
```

In[7]:
```
titanic_data = pd.concat([titanic_data,
                          pd.get_dummies(titanic_data['Sex']),
                          pd.get_dummies(titanic_data['Embarked'],prefix="embark"),
                          pd.get_dummies(titanic_data['Pclass'],prefix="class")], axis=1)
```

```python
titanic_data["Age"] = titanic_data["Age"].fillna(titanic_data["Age"].mean())
titanic_data["Fare"] = titanic_data["Fare"].fillna(titanic_data["Fare"].mean())
titanic_data = titanic_data.drop(['Name','Ticket','Cabin','Sex','Embarked',
'Pclass'], axis=1)
```

In[8]:
```python
# 设置固定的随机种子以确保可复现
np.random.seed(707)
# 将特征和标签转换为 numpy 数组
labels = titanic_data["Survived"].to_numpy()
titanic_data = titanic_data.drop(['Survived'], axis=1)
feature_names = list(titanic_data.columns)
data = titanic_data.to_numpy()
# 将训练集和测试集进行分离
train_indices = np.random.choice(len(labels), int(0.7*len(labels)),
replace=False)
test_indices = list(set(range(len(labels))) - set(train_indices))
train_features = data[train_indices]
train_labels = labels[train_indices]
test_features = data[test_indices]
test_labels = labels[test_indices]
```

In[9]:
```python
train_features.shape
```

Out[9]:

(623, 12)

由上述代码分析可知，数据集中有 623 条记录可供训练模型，每条记录有 12 个特征可用于模型的学习。通过使用 Torch 的神经网络模块，首先在第一隐藏层设计了 12 个隐藏神经元，接着在第二隐藏层设计了另外 12 个神经元。最终，目标变量设置两个标签作为输出的结果。在这两个隐藏层中，均采用了线性结构并使用了 Sigmoid 激活函数。在最后一层，采用了 Softmax 激活函数，以便得到类别概率。

In[10]:
```python
import torch
import torch.nn as nn
```

```python
torch.manual_seed(1)  # 设置固定的随机种子以确保可复现
class TitanicSimpleNNModel(nn.Module):
    def __init__(self):
        super().__init__()
        self.linear1 = nn.Linear(12, 12)
        self.sigmoid1 = nn.Sigmoid()
        self.linear2 = nn.Linear(12, 8)
        self.sigmoid2 = nn.Sigmoid()
        self.linear3 = nn.Linear(8, 2)
        self.softmax = nn.Softmax(dim=1)
    def forward(self, x):
        lin1_out = self.linear1(x)
        sigmoid_out1 = self.sigmoid1(lin1_out)
        sigmoid_out2 = self.sigmoid2(self.linear2(sigmoid_out1))
        return self.softmax(self.linear3(sigmoid_out2))
```

In [11]:

```python
net = TitanicSimpleNNModel()
criterion = nn.CrossEntropyLoss()
num_epochs = 200

optimizer = torch.optim.Adam(net.parameters(), lr=0.1)
input_tensor = torch.from_numpy(train_features).type(torch.FloatTensor)
label_tensor = torch.from_numpy(train_labels)
```

在模型之中，我们选用交叉熵损失函数作为损失函数。根据输入数据及所需精度，在实际项目中，可便捷地选用不同类型的损失函数。此外，鉴于 Adam 优化器在多数场景下的适用性，该模型也采用了 Adam 优化器进行优化。

In [12]:

```python
for epoch in range(num_epochs):
    output = net(input_tensor)
    loss = criterion(output, label_tensor)
    optimizer.zero_grad()
    loss.backward()
    optimizer.step()
    if epoch % 20 == 0:
        print ('Epoch {}/{} => Loss: {:.2f}'.format(epoch+1, num_epochs,
        loss.item()))
torch.save(net.state_dict(), '/model.pt')
```

```
Epoch 1/200 => Loss: 0.70
Epoch 21/200 => Loss: 0.55
Epoch 41/200 => Loss: 0.50
Epoch 61/200 => Loss: 0.49
Epoch 81/200 => Loss: 0.48
Epoch 101/200 => Loss: 0.49
Epoch 121/200 => Loss: 0.47
Epoch 141/200 => Loss: 0.47
Epoch 161/200 => Loss: 0.47
Epoch 181/200 => Loss: 0.47
```

在上述代码中，为了尽可能降低交叉熵损失，进行了 200 次迭代。通过调用 torch.save 函数，将训练好的模型保存至默认目录。根据实际需求，可以调整保存模型的路径，默认路径为 /model。

In [13]:
```
out_probs = net(input_tensor).detach().numpy()
out_classes = np.argmax(out_probs, axis=1)
print("Train Accuracy:", sum(out_classes == train_labels) / len(train_labels))
Train Accuracy: 0.8523274478330658
```

为了储存类别概率，输入的张量从神经网络模型中分离，并转化成了一个 Numpy 数组。训练准确率为 85.23%。

In [14]:
```
test_input_tensor = torch.from_numpy(test_features).type(torch.FloatTensor)
out_probs = net(test_input_tensor).detach().numpy()
out_classes = np.argmax(out_probs, axis=1)
print("Test Accuracy:", sum(out_classes == test_labels) / len(test_labels))

Test Accuracy: 0.832089552238806
```

集成梯度（IG）可以从神经网络模型中进行提取。通过调用 attribute( ) 函数，可以实现集成梯度的提取。在此过程中，需要将 return_convergence_delta 参数设置为 True，并在测试数据集上应用 requires_grad_( ) 方法。

In [15]:
```
ig = IntegratedGradients(net)
```

In [16]:
```
test_input_tensor.requires_grad_()
attr, delta = ig.attribute(test_input_tensor,target=1, return_convergence_
```

```
delta=True)
attr = attr.detach().numpy()
```

In [17]:
```
np.round(attr,2)
```

Out[17]:
```
array([[-0.7 , 0.09, -0. , ..., 0. , 0. , -0.33], [-2.78, -0. , -0. , ..., 0. , 0. , -1.82], [-0.65, 0. , -0. , ..., 0. , 0. , -0.31], ..., [-0.47, -0. , -0. , ..., 0.71, 0. , -0. ], [-0.1 , -0. , -0. , ..., 0. , 0. , -0.1 ], [-0.7 , 0. , -0. , ..., 0. , 0. , -0.28]])
```

变量 attr 包含了模型对输入特征的重要性的度量。在代码中，使用 IntegratedGradients 方法对输入张量进行归因操作，将目标设置为 1，并通过设置 return_convergence_delta 参数为 True 来获取归因结果和收敛差值。归因结果存储在变量"attr"中，通过将其转换为 NumPy 数组，可以看到每个输入特征的重要性值。这些重要性值表示了模型对输入特征的影响程度，较高的值意味着该特征对模型的输出结果具有更大的影响。

In [18]:
```
importances = np.mean(attr, axis=0)
```

In [19]:
```
for i in range(len(feature_names)):
        print(feature_names[i], ": ", '%.3f'%(importances[i]))
Age :  -0.574
SibSp :  -0.010
Parch :  -0.026
Fare :  0.278
female :  0.101
male :  -0.460
embark_C :  0.042
embark_Q :  0.005
embark_S :  -0.021
class_1 :  0.067
class_2 :  0.090
class_3 :  -0.144
```

特征重要性可以是负数或正数，其具体含义如下：负数表示该特征对分类概率产生负面影响，即减少分类概率得分；而正数则表示增加分类概率得分，即增加该分类的概率。因此，特征重要性能够体现特征在进行分类时的相关性。

秘籍 10-3 深度学习模型中神经元的重要性

问题

如何计算深度学习模型中神经元的重要性？

解决方案

导纳层通过计算神经元相对于输入和输出的偏导数，对神经元进行激活组合，以评估神经元的重要性。这一层基于集成梯度构建，通过分析集成梯度归因的流动，以确定神经元在网络中的重要性。

编程实战

下面这段代码展示了如何计算深度学习模型中神经元的重要性。

```
In[20]:
cond = LayerConductance(net, net.sigmoid1)
```

上面这行代码中，net.sigmoid1 代表深度学习模型 net 的第一个隐藏层。在此，net 为深度学习模型对象，而 LayerConductance 则表示导纳层函数。计算得到的导纳值（conductance）将被保存在 cond 变量中。

```
In[21]:
cond_vals = cond.attribute(test_input_tensor,target=1)
cond_vals = cond_vals.detach().numpy()
In[22]:
Average_Neuron_Importances = np.mean(cond_vals, axis=0)
Average_Neuron_Importances
Out[22]:
array([ 0.03051018, -0.23244175,  0.04743345,  0.02102091, -0.08071412,
```

-0.09040915, -0.13398956, -0.04666219, 0.03577907, -0.07206058,
-0.15658873, 0.03491106], dtype=float32)

隐藏层设有 12 个神经元，故神经元重要性列表相应包含 12 个项目。考虑到存在多个隐藏层，因此取其导纳值的平均值作为最终结果。

In [23]:

neuron_cond = NeuronConductance(net, net.sigmoid1)

相应地，NeuronConductance 函数用于提取隐藏层神经元的导纳值。

In [24]:

neuron_cond_vals_10 = neuron_cond.attribute(test_input_tensor, neuron_selector=10, target=1)

In [25]:

neuron_cond_vals_0 = neuron_cond.attribute(test_input_tensor, neuron_selector=0, target=1)

In [26]:

# 第 0 个神经元的平均特征重要性
nn0 = neuron_cond_vals_0.mean(dim=0).detach().numpy()
np.round(nn0,3)

Out[26]:
array([0.008, 0. , 0. , 0.028, 0. , -0.004, -0. , 0. , -0.001, -0. , 0. ,
-0.], dtype=float32)

秘籍 10-4 安装 Skorch 库

问题

如何进行 Skorch 的安装？

解决方案

安装 Skorch 有两种途径：一是通过 conda 指令，二是通过 pip 安装。

编程实战

可以按照以下步骤安装该库。

In[27]:
pip install -U skorch

或

conda install -U skorch

Out[27]:
```
Looking in indexes: https://pypi.org/simple, https://us-python.pkg.dev/colab-wheels/public/simple/
Collecting skorch
  Downloading skorch-0.11.0-py3-none-any.whl (155 kB)
     |████████████████████████████████| 155 kB 27.9 MB/s
Requirement already satisfied: numpy>=1.13.3 in /usr/local/lib/python3.7/dist-packages (from skorch) (1.21.6)
Requirement already satisfied: scikit-learn>=0.19.1 in /usr/local/lib/python3.7/dist-packages (from skorch) (1.0.2)
Requirement already satisfied: tqdm>=4.14.0 in /usr/local/lib/python3.7/dist-packages (from skorch) (4.64.0)
Requirement already satisfied: scipy>=1.1.0 in /usr/local/lib/python3.7/dist-packages (from skorch) (1.7.3)
Requirement already satisfied: tabulate>=0.7.7 in /usr/local/lib/python3.7/dist-packages (from skorch) (0.8.10)
Requirement already satisfied: joblib>=0.11 in /usr/local/lib/python3.7/dist-packages (from scikit-learn>=0.19.1->skorch) (1.1.0)
Requirement already satisfied: threadpoolctl>=2.0.0 in /usr/local/lib/python3.7/dist-packages (from scikit-learn>=0.19.1->skorch) (3.1.0)
Installing collected packages: skorch
Successfully installed skorch-0.11.0
```

秘籍 10-5　Skorch 组件在神经网络分类器中的应用

问题

如何训练基于 Skorch 的神经网络分类器？

解决方案

Skorch 是一款与 Scikit-learn 兼容的神经网络库，封装了 PyTorch，为用户提供了便捷的神经网络训练功能，有效降低了冗余代码的编写需求。该库适用于分类和回归任务，主要模块 skorch.NeuralNetClassifier 和 skorch.NeuralNetRegressor 具备优异性能，训练过程迅速，结果展示美观。

编程实战

Scikit-learn 的诸多组件如拟合、预处理、预测、交叉验证、度量指标、网格搜索以及流水线在业界广受欢迎。而 PyTorch 已逐渐成为训练各类深度学习模型的首选工具。因此，我们期望在 PyTorch 库所训练的神经网络模型中，运用 Scikit-learn 的这些优质组件。

```
In[28]:
import torch
from torch import nn
import numpy as np
import torch.nn.functional as F
from sklearn.datasets import make_classification
In[29]:
X, y = make_classification(2000, 10, random_state=0)
X, y = X.astype(np.float32), y.astype(np.int64)
```

以上代码包含用于分类任务的标准示例数据。

In [30]:
```python
class ClassifierModule(nn.Module):
    def __init__(
            self,
            num_units=30,
            nonlin=F.relu,
            dropout=0.5,
    ):
        super(ClassifierModule, self).__init__()
        self.num_units = num_units
        self.nonlin = nonlin
        self.dropout = dropout

        self.dense0 = nn.Linear(10, num_units)
        self.nonlin = nonlin
        self.dropout = nn.Dropout(dropout)
        self.dense1 = nn.Linear(num_units, 10)
        self.output = nn.Linear(10, 2)

    def forward(self, X, **kwargs):
        X = self.nonlin(self.dense0(X))
        X = self.dropout(X)
        X = F.relu(self.dense1(X))
        X = F.softmax(self.output(X), dim=-1)
        return X
```

In [31]:
```python
from skorch import NeuralNetClassifier
```

In [32]:
```python
net = NeuralNetClassifier(
    ClassifierModule,
    max_epochs=20,
    lr=0.1,
#   device='cuda',   # 取消此行注释以使用 CUDA 进行训练
)
```

如果运行设备具备 GPU 环境，则可以将上述分类器中的注释代码取消注释。

In [33]:
```python
net.get_params()
```

In[34]:
net.fit(X, y)

epoch	train_loss	valid_acc	valid_loss	dur
1	0.7076	0.5325	0.6674	0.0255
2	0.6532	0.8050	0.5975	0.0217
3	0.5660	0.9675	0.4638	0.0206
4	0.4265	0.9800	0.2979	0.0189
5	0.2913	0.9875	0.1789	0.0207
6	0.2128	0.9925	0.1150	0.0208
7	0.1689	0.9900	0.0825	0.0207
8	0.1496	0.9900	0.0644	0.0195
9	0.1183	0.9900	0.0545	0.0197
10	0.1218	0.9900	0.0490	0.0198
11	0.1240	0.9900	0.0464	0.0203
12	0.1090	0.9900	0.0428	0.0213
13	0.1050	0.9900	0.0410	0.0215
14	0.1067	0.9875	0.0399	0.0214
15	0.1072	0.9900	0.0392	0.0211
16	0.0958	0.9900	0.0378	0.0241
17	0.0964	0.9900	0.0371	0.0213
18	0.0986	0.9925	0.0361	0.0223
19	0.0884	0.9900	0.0363	0.0205
20	0.0991	0.9900	0.0366	0.0213

Out[34]:
```
<class 'skorch.classifier.NeuralNetClassifier'>[initialized](
  module_=ClassifierModule(
    (dense0): Linear(in_features=10, out_features=30, bias=True)
    (dropout): Dropout(p=0.5, inplace=False)
    (dense1): Linear(in_features=30, out_features=10, bias=True)
    (output): Linear(in_features=10, out_features=2, bias=True)
  ),
)
```

In[35]:
list(net.get_params())

Out[35]:
['module', 'criterion', 'optimizer', 'lr', 'max_epochs', 'batch_

size', 'iterator_train', 'iterator_valid', 'dataset', 'train_split', 'callbacks', 'predict_nonlinearity', 'warm_start', 'verbose', 'device', '_kwargs', 'classes', 'callbacks__epoch_timer', 'callbacks__train_loss', 'callbacks__train_loss__name', 'callbacks__train_loss__lower_is_better', 'callbacks__train_loss__on_train', 'callbacks__valid_loss', 'callbacks__valid_loss__name', 'callbacks__valid_loss__lower_is_better', 'callbacks__valid_loss__on_train', 'callbacks__valid_acc', 'callbacks__valid_acc__scoring', 'callbacks__valid_acc__lower_is_better', 'callbacks__valid_acc__on_train', 'callbacks__valid_acc__name', 'callbacks__valid_acc__target_extractor', 'callbacks__valid_acc__use_caching', 'callbacks__print_log', 'callbacks__print_log__keys_ignored', 'callbacks__print_log__sink', 'callbacks__print_log__tablefmt', 'callbacks__print_log__floatfmt', 'callbacks__print_log__stralign'])

在模型训练完成后，可以运用 Scikit-learn 库中的 predict 函数来生成预测结果，并可通过 predict_proba 函数获取各个类别的概率分布。

In[36]:

```
y_pred = net.predict(X[:5])
y_pred
```

Out[36]:

```
array([1, 0, 0, 1, 1])
```

In[37]:

```
y_proba = net.predict_proba(X[:5])
y_proba
```

Out[37]:

```
array([[7.7738642e-04, 9.9922264e-01], [9.9628782e-01, 3.7122301e-03],
[9.9648917e-01, 3.5108225e-03], [3.2411060e-01, 6.7588937e-01],
[4.5940662e-03, 9.9540591e-01]], dtype=float32)
```

秘籍 10-6 Skorch 神经网络回归器

问题

如何使用 Skorch 训练回归模型？

解决方案

遵循神经网络模型的常规方法，我们对回归模型进行训练。为此，采用一个生成合成数据的回归器函数，并利用 Skorch 库的相应函数来训练模型。

编程实战

以下代码示例展示了如何实现该功能。

```
In[38]:
from sklearn.datasets import make_regression
In[39]:
X_regr, y_regr = make_regression(1000, 20, n_informative=10,
random_state=0)
X_regr = X_regr.astype(np.float32)
y_regr = y_regr.astype(np.float32) / 100
y_regr = y_regr.reshape(-1, 1)
In[40]:
X_regr.shape, y_regr.shape, y_regr.min(), y_regr.max()
Out[40]:
((1000, 20), (1000, 1), -6.4901485, 6.154505)
In[41]:
class RegressorModule(nn.Module):
    def __init__(
            self,
            num_units=10,
            nonlin=F.relu,
    ):
        super(RegressorModule, self).__init__()
        self.num_units = num_units
        self.nonlin = nonlin

        self.dense0 = nn.Linear(20, num_units)
        self.nonlin = nonlin
```

```python
        self.dense1 = nn.Linear(num_units, 10)
        self.output = nn.Linear(10, 1)

    def forward(self, X, **kwargs):
        X = self.nonlin(self.dense0(X))
        X = F.relu(self.dense1(X))
        X = self.output(X)
        return X
```

In[42]:

```python
from skorch import NeuralNetRegressor
```

In[43]:

```python
net_regr = NeuralNetRegressor(
    RegressorModule,
    max_epochs=20,
    lr=0.1,
#   device='cuda',  # 取消此行注释以使用 CUDA 进行训练
)
```

In[44]:

```python
net_regr.fit(X_regr, y_regr)
```

epoch	train_loss	valid_loss	dur
1	4.3247	3.0078	0.0170
2	1.7262	0.6808	0.0123
3	0.6510	0.2147	0.0115
4	0.1811	0.2132	0.0118
5	0.1906	0.1127	0.0108
6	0.1143	0.3361	0.0204
7	0.3835	0.0899	0.0113
8	0.0845	0.1574	0.0117
9	0.1099	0.0486	0.0130
10	0.0485	0.0974	0.0128
11	0.0907	0.0447	0.0108
12	0.0481	0.0947	0.0129
13	0.0881	0.0322	0.0128
14	0.0323	0.0599	0.0117
15	0.0461	0.0180	0.0115
16	0.0161	0.0328	0.0123

17	0.0231	0.0125	0.0123
18	0.0098	0.0208	0.0123
19	0.0143	0.0102	0.0112
20	0.0074	0.0153	0.0121

Out[44]:

```
<class 'skorch.regressor.NeuralNetRegressor'>[initialized](
  module_=RegressorModule(
    (dense0): Linear(in_features=20, out_features=10, bias=True)
    (dense1): Linear(in_features=10, out_features=10, bias=True)
    (output): Linear(in_features=10, out_features=1, bias=True)
  ),
)
```

In[45]:

```
y_pred = net_regr.predict(X_regr[:5])
y_pred
```

Out[45]:

```
array([[ 0.7368696 ], [-1.2884711 ], [-0.51758516], [-0.11890286],
[-0.61254007]], dtype=float32)
```

  在此示例中，我们展示了如何使用 Skorch 库训练一个回归模型。包括数据准备、模型定义、模型训练和模型预测。

  在**数据准备**阶段，使用 make_regression 函数生成了一个包含 1000 个样本和 20 个特征的合成数据集。

  在**模型定义**阶段，定义了一个名为 RegressorModule 的神经网络模型类，继承自 nn.Module。该模型类包含了几个线性层和激活函数，用于构建回归模型的前向传播过程。

  在**模型训练**阶段，我们先导入 NeuralNetRegressor 类，该类是 Skorch 中用于回归任务的神经网络回归器。接着，创建了一个 NeuralNetRegressor 对象 net_regr，传入了之前定义的 RegressorModule 作为模型参数，并设置最大训练轮数和学习率。然后，调用 fit 方法，将训练数据 X_regr 和目标变量 y_regr 传入模型进行训练。训练过程显示了每个训练轮次的训练损失和验证损失。

  最后，在**模型预测**阶段，我们使用训练好的模型进行预测，通过调用 predict 方法传入输入数据 X_regr[:5]，得到了对应的回归预测结果。

秘籍 10-7　Skorch 模型的保存和加载

问题

如何保存和加载由 Skorch 生成的模型对象？

解决方案

使用 pickle 库，可以将模型对象保存为序列化的形式，然后将其加载到另一个环境中。本节将演示如何保存和加载 Skorch 的模型。

编程实战

以下代码展示了如何执行此操作。

In[46]:
```
import pickle
```
In[47]:
```
file_name = '/tmp/mymodel.pkl'
```
In[48]:
```
with open(file_name, 'wb') as f:
    pickle.dump(net, f)
```
In[49]:
```
with open(file_name, 'rb') as f:
    new_net = pickle.load(f)
```
In[50]:
```
net.save_params(f_params=file_name)  # 此处也可以使用文件处理程序
```

如果将模型对象存储为已保存的参数，则需要再次初始化模型，并将其分配给一个新对象。

In[51]:
# 首先初始化模型
new_net = NeuralNetClassifier(
    ClassifierModule,
    max_epochs=20,
    lr=0.1,
).initialize()

In[52]:
new_net.load_params(file_name)

秘籍 10-8　使用 Skorch 创建神经网络模型流水线

问题

如何使用 Skorch 为神经网络模型创建流水线？

解决方案

流水线对象作为一种结构，具备按预定顺序组织和执行一系列操作的能力。在模型训练与执行过程中，这种方式确保各项操作遵循指定顺序，从而实现模型的有序执行。

编程实战

以下脚本展示了如何使用 Skorch 实现这一点。

In[53]:
```
from sklearn.pipeline import Pipeline
from sklearn.preprocessing import StandardScaler
```

In[54]:
```
pipe = Pipeline([
    ('scale', StandardScaler()),
```

```
    ('net', net),
])
```

In[55]:

```
pipe.fit(X, y)
Re-initializing module.
Re-initializing criterion.
Re-initializing optimizer.
```

epoch	train_loss	valid_acc	valid_loss	dur
1	0.6925	0.5250	0.6640	0.0355
2	0.6361	0.9075	0.5834	0.0354
3	0.5447	0.9550	0.4427	0.0340
4	0.4197	0.9675	0.2898	0.0207
5	0.3019	0.9775	0.1798	0.0214
6	0.2282	0.9825	0.1206	0.0217
7	0.1790	0.9875	0.0869	0.0207
8	0.1550	0.9875	0.0697	0.0208
9	0.1473	0.9875	0.0594	0.0196
10	0.1249	0.9875	0.0525	0.0200
11	0.1294	0.9900	0.0482	0.0213
12	0.1194	0.9925	0.0446	0.0220
13	0.1192	0.9950	0.0428	0.0291
14	0.1035	0.9925	0.0406	0.0215
15	0.0989	0.9925	0.0394	0.0223
16	0.0999	0.9925	0.0386	0.0206
17	0.0928	0.9925	0.0376	0.0200
18	0.0980	0.9925	0.0370	0.0189
19	0.0969	0.9925	0.0364	0.0211
20	0.0876	0.9925	0.0358	0.0213

Out[55]:

```
Pipeline(steps=[('scale', StandardScaler()), ('net', <class 'skorch.
classifier.NeuralNetClassifier'>[initialized]( module_=ClassifierModule
( (dense0): Linear(in_features=10, out_features=30, bias=True) (dropout):
Dropout(p=0.5, inplace=False) (dense1): Linear(in_features=30, out_
features=10, bias=True) (output): Linear(in_features=10, out_features=2,
bias=True) ), ))])
```

In[56]:
```
y_proba = pipe.predict_proba(X[:5])
y_proba
```
Out[56]:
```
array([[0.00224374, 0.9977563 ], [0.9986193 , 0.00138069], [0.99899906,
0.00100095], [0.30393705, 0.6960629 ], [0.00816792, 0.9918321 ]],
dtype=float32)
```

秘籍 10-9　使用 Skorch 进行神经网络模型的轮次评分

问题

如何在使用 Skorch 训练神经网络模型时使用回调函数，用于对预测结果进行通用评分？

解决方案

在训练深度学习模型时，可以利用回调函数进行轮次评分。因此，需要定义一个评分函数。在每个轮次完成后，需要调用该函数，且如果达到了期望的准确度水平，则应该进行高亮显示。

编程实战

以下代码展示了实现方法。

In[57]:
```
from skorch.callbacks import EpochScoring
```
In[58]:
```
auc = EpochScoring(scoring='roc_auc', lower_is_better=False)
```

In[59]:
```
net = NeuralNetClassifier(
    ClassifierModule,
    max_epochs=20,
    lr=0.1,
    callbacks=[auc],
)
```
In[60]:
```
net.fit(X, y)
```

epoch	roc_auc	train_loss	valid_acc	valid_loss	dur
1	0.9544	0.6614	0.8900	0.6294	0.0193
2	0.9845	0.5875	0.9625	0.5233	0.0208
3	0.9899	0.4798	0.9800	0.3647	0.0198
4	0.9945	0.3600	0.9825	0.2302	0.0208
5	0.9972	0.2682	0.9850	0.1451	0.0200
6	0.9975	0.2087	0.9850	0.1002	0.0187
7	0.9978	0.1869	0.9850	0.0762	0.0191
8	0.9979	0.1699	0.9850	0.0640	0.0222
9	0.9980	0.1430	0.9875	0.0567	0.0200
10	0.9981	0.1338	0.9875	0.0500	0.0201
11	0.9981	0.1214	0.9875	0.0464	0.0355
12	0.9981	0.1167	0.9900	0.0442	0.0346
13	0.9982	0.1072	0.9875	0.0419	0.0396
14	0.9981	0.1152	0.9900	0.0404	0.0337
15	0.9981	0.1086	0.9900	0.0395	0.0341
16	0.9982	0.0905	0.9875	0.0387	0.0338
17	0.9981	0.0983	0.9875	0.0382	0.0466
18	0.9981	0.0929	0.9875	0.0373	0.0618
19	0.9982	0.1009	0.9875	0.0368	0.0302
20	0.9982	0.0981	0.9875	0.0362	0.0219

Out[60]:
<class 'skorch.classifier.NeuralNetClassifier'>[initialized]
(module_=ClassifierModule((dense0): Linear(in_features=10, out_features=30, bias=True) (dropout): Dropout(p=0.5, inplace=False) (dense1): Linear(in_features=30, out_features=10, bias=True) (output): Linear(in_features=10, out_features=2, bias=True)),)

In[61]:
```
print(', '.join(net.prefixes_))
iterator_train, iterator_valid, callbacks, dataset, module, criterion, optimizer
```

秘籍 10-10　使用 Skorch 进行超参数的网格搜索

问题

如何使用 Skorch 进行超参数训练的网格搜索？

解决方案

在深度学习模型中，不同的超参数组合可以生成多个模型。为了确定哪些超参数组合能够产生最优模型，则需要采用一种特定的方法。这些能够生成最优模型的超参数组合被定义为最佳超参数。值得注意的是，随着设定最大训练轮次的增加，最佳超参数可能会发生变化。

编程实战

以下脚本展示了实现方法。

In[62]:
```
from sklearn.model_selection import GridSearchCV
```

In[63]:
```
net = NeuralNetClassifier(
    ClassifierModule,
    max_epochs=20,
    lr=0.1,
    optimizer__momentum=0.9,
    verbose=0,
    train_split=False,
)
```

In[64]:
```python
params = {
    'lr': [0.05, 0.1],
    'module__num_units': [10, 20],
    'module__dropout': [0, 0.5],
    'optimizer__nesterov': [False, True],
}
```

In[65]:
```python
gs = GridSearchCV(net, params, refit=False, cv=3, scoring='accuracy', verbose=2)
```

In[66]:
```
gs.fit(X, y)
Fitting 3 folds for each of 16 candidates, totalling 48 fits
[CV] END lr=0.05, module__dropout=0, module__num_units=10, optimizer__nesterov=False; total time=   0.4s
[CV] END lr=0.05, module__dropout=0, module__num_units=10, optimizer__nesterov=False; total time=   0.3s
[CV] END lr=0.05, module__dropout=0, module__num_units=10, optimizer__nesterov=False; total time=   0.3s
[CV] END lr=0.05, module__dropout=0, module__num_units=10, optimizer__nesterov=True; total time=   0.3s
[CV] END lr=0.05, module__dropout=0, module__num_units=10, optimizer__nesterov=True; total time=   0.3s
[CV] END lr=0.05, module__dropout=0, module__num_units=10, optimizer__nesterov=True; total time=   0.3s
[CV] END lr=0.05, module__dropout=0, module__num_units=20, optimizer__nesterov=False; total time=   0.3s
[CV] END lr=0.05, module__dropout=0, module__num_units=20, optimizer__nesterov=False; total time=   0.3s
[CV] END lr=0.05, module__dropout=0, module__num_units=20, optimizer__nesterov=False; total time=   0.3s
[CV] END lr=0.05, module__dropout=0, module__num_units=20, optimizer__nesterov=True; total time=   0.3s
[CV] END lr=0.05, module__dropout=0, module__num_units=20, optimizer__nesterov=True; total time=   0.3s
[CV] END lr=0.05, module__dropout=0, module__num_units=20, optimizer__nesterov=True; total time=   0.3s
```

```
[CV] END lr=0.05, module__dropout=0.5, module__num_units=10, optimizer__nesterov=False; total time=   0.3s
[CV] END lr=0.05, module__dropout=0.5, module__num_units=10, optim..................
```

In [67]:
```
print(gs.best_score_, gs.best_params_)
.988499744121933 {'lr': 0.1, 'module__dropout': 0.5, 'module__num_units': 20, 'optimizer__nesterov': False}
```

小结

  本章探讨了Skorch库，这是一个与Scikit-learn兼容的神经网络库，它封装了PyTorch，为用户提供了神经网络训练的便捷功能。此外，本章还特别介绍了模型可解释性的应用方法，这对于解决各种监督学习相关任务（如回归和分类）的深度学习模型来说，具有至关重要的意义。